I0475310

Math 4 2-Day

Math 4 2-Day
Tips, Tricks, Topics and Techniques in Everyday Mathematics

Michael R. Irwin

Cincinnati
2011

Copyright © 2006-2011 Michael R. Irwin
All rights reserved under International and Pan-American
Copyright Conventions.
Fair usage laws apply under International and U. S. Copyright Laws

ISBN-13: 978-1460993026
ISBN-10: 1460993020

[Dedication]

This book is dedicated to all my children – Richard, Joseph, and David Irwin

- Table of Contents -

Preface i

Introduction iii

Acknowledgements/Credits iv

Day One 1

Quotations of the Day 2

Bit of Fun in Math
10 Commandments (or concepts) of Math 3
7 times 13 always equals 28! 4

Review of Types of Numbers
What are Numbers? 5

Numbers (many different Sorts and Types)
Natural Numbers (or the Counting Numbers)...6
Whole Numbers 6
Integers...8
Rational Numbers 8
Irrational Numbers 9
Transcendental and Algebraic Numbers 9

Avoiding Careless Error Math (Part 1)
Definitions.. 10
Casting Out Nines 12
Using Check Digits in Addition Problems.. 12

Simple Math is All You Need (Part 1)
Some Basics of Simple Math 14
Multiplication Fun with 10, 5, and 2 15
Fun with multiplying by 11 17
Multiplying any two digit number by another two digit number 21
Multiplying by powers of 2 – 2, 4, 8 22

Calculator Tricks (Part 1)
Is that your final answer?.. 23
The Birthday display.. 23
Cheat the Calculator 23

Bits and Pieces of Algebra
The magic number 5 (also known as ALL THE SAME) 24
Reading your mind 24

Endgame for the Day
Get off the Earth puzzle! 25

Day Two 27

Quotations of the Day 28

Bit of Fun in Math
I have a friend who 29
Land of Candy.. 29

Five Easy Ideas in Math
    Counting PARTS up and down 30
    Work with compatible numbers first.. 31
    Performing equal addition to solve subtraction problems 31
    Breaking apart problems and numbers... 32
    Using Compensation 33

Avoiding Careless Error Math (Part 2)
    Review: Using Check digits in Addition 35
    Using Check Digits in Subtraction Problems... 35
    Subtraction and Check Digits 36

Mathematical Curiosities (Part 1)
    The Exciting World of Palindrome Numbers.. 38
    Interesting Palindrome facts... 39

Simple Math is All You Need (Part 2)
    Multiplying any number of digits by any number of 9's.................................. 40
    Revisit multiplying any number by 2 or any power of 2 (2^2, 2^3, ...)............ 42
    Revisit multiplying any number by 5 42

A few Number Oddities
    The Apocalyptic Number 666 43
    Is a Billion, a billion through the World? 43
    Cyclic numbers in the world of mathematics... 44
    The ROMAN problem of XI + I = X or 11 + 1 = 10 45

Calculator Tricks (Part 2)
    The Count is IN! *or* It's all about the zeros and a little old decimal 46
    Words, Words, Words (up and over) 46
    Not quite 1,000,000 (start with one less) 46

Math is Logic – or – the Logical Methods of Math (Part 1)
    Reasoning out word problems.. 47
    What is logic? 48
    A few 'easy' logic problems.. 49
    Strategies for solving logic problems 49

End Game for the Day
    Creating a money shirt 52

Day Three **55**

Quotations of the Day..56

Bit of Fun in Math
    Murphy's Law in Mathematics.. 57

Laws of Math
    Mathematical Properties 58
    Order of Precedence... 61

Mathematical Curiosities (Part 2)
    Prime Numbers in the world of math.. 65
    Sieve of Eratosthenes.. 66

Fractions – the math of parts
    Least Common Denominator and Greatest Common Factor 67
    Upside-down Birthday Cake method for finding GCF and LCD...................... 70
    Venn diagram method for finding GCF and LCD 73

What are Fractions?...77
Adding Fractions 77
Subtracting fractions...78
Multiplying with Fractions 79
Dividing with Fractions 80

Calculator Tricks (Part 3)
Good Luck or Bad Luck 81
Secret of 73 81
Phone Number Trick 81

Math is Logic – or – The Logical Methods of Math (Part 2)
Puzzle 1: Crates of Fruit 82
Puzzle 2: Matching Socks 82
Puzzle 3: Relative Picture...83
Puzzle 4: The Boy and Girl...83
Puzzle 5: My Sons 83
Puzzle 6: Twin Brothers 83

Simple Math is All You Need (Part 3)
Multiplying any number times 12..84
Multiplying two numbers less than and near 100...85
Squaring any number ending in 0..86
Squaring any number ending in 5..86
Another rapid method for multiplying any number less than 100 by 99 87
Rapid Multiply Two digits times 101 87
Rapid Multiply Three Digits by 1001 88
Multiplying two teen numbers that differ by 1..88

Endgame for the Day
Connect the dots..89

Day Four **91**

Quotations of the Day 92

Bit of Fun in Math
Math over the Years...93

Mathematical Curiosities – (Part 3)
Enormous numbers and ways people visualize them..94
Using scientific notation to handle large and small numbers............................95
Large Numbers by 10 to the power of:...97
The Googol family 97
Other large numbers used in mathematics..98

Math is Logic – or – The Logical Methods of Math (Part 3)
Probability is 102
Formula for the Probability of an Event 104

Curious methods for Multiplication
History of multiplication 107
Curious methods of Multiplication 109

Calculator Tricks – (Part 4)
The 6174 loop 115
Another Birthday display 115
The Mind Reading Number Trick...115

Simple Math is All You Need – (Part 4)
Determine if a number is even divisible by 2 through 10 116
Multiplying any same number digits by same number of 9's................................117
Summing a Sequential Series of Numbers...118

Endgame for the Day
The Puzzle: 5 lines of 10 balls 119
Cupids Arrow 119

Day Five...**121**

Quotations of the Day.. 122

Bit of Fun in Math
How much is 2 times 2?..123
EDUCATION = Problems!..124

Mathematical Curiosities – (Part 4)
The Fibonacci Series 125
Fibonacci Series in Nature..128
Fibonacci Fingers?..132
Fibonacci and the Golden Ratio..132
The Golden Ratio/Golden Section or Phi Φ 132
The Human body and the Golden Section 134
Divine Proportion and Art...135
Architecture and Fibonacci 137

Avoiding Careless Error Math (Part 3)
Verifying Addition and Subtraction problems using Check digits 138
Using Check Digits in Multiplication..139
Using check digits in multiplying...139
Using Check Digits to fix an Error in Multiplication 140
MOD (7) Residue Double Check Method 142
Understanding the MOD 7 residue method 143

Computers and the Math behind Them
A Short History of Numbers and Numeral Systems..145
Base-2 to the Rescue! 147

Computers and Standard ASCII Character Set..151

Calculator Tricks – (Part 5)
The Golden Prediction 153
The 4-2-1 Loop 153
Mystery of 24 and Prime Numbers 153

Simple Math is All You Need – (Part 5)
Multiply any number by 10...156
Multiply any number by 2..156
Multiply any number by 5 156
Mix it Up! 157
Squaring any number ONE below or above a previous square: 158
Squaring any number TWO below or above a previous square:...............................158
Squaring any number quickly between 10 and 120 158
Even Divisibility by 11...159
Rapid Addition using 'Process of 11' 159

Endgame for the Day
    What is an Optical Illusion?.. 160
    Mathematics and Optical Illusions 160
    A few Optical Illusions 161

Day Six..**167**

Quotations of the Day...168

Bit of Fun in Math
    There are 1 0 types of people in the world... 169
    Top 10 excuses for not doing your homework!.. 170

Mathematical Curiosities – (Part 5)
    Möbius and his impossible shapes!.. 171
    A four dimensional (4D) Möbius object.. 173
    A Möbius card 176

Exponentiation or Raising to the Power of
    What is Exponentiation 177
    Rules of Exponentiation and Algebra 178

Calculator Tricks –(Part 6)
    Superfast Addition 179
    Doubling three digits – again and again 179
    MacArthur's magic 115 179

Simple Math is All You Need – (Part 6)
    Converting Fahrenheit to Celsius and back... 180
    A Grab Bag of Methods 183
    Fun with numbers and oddities.. 186
    The beauty of symmetrical math.. 187

Endgame for the Day
    Doing Ancient Math... 180

Appendix **191**

Author Guide and Comments..191

Subtraction through Addition
    Using Addition while Subtracting.. 193

School Bus Problem – What Direction 195

Two Neurological Tests
    A Test of 'F's 196
    Finding the 6 in a bed of 9s 196

Thinking Algebraically
    Cost of a Baseball and Glove 197

M, Heart, 8 Problem 198

INDEX **201**

Preface

THIS BOOK CONTAINS THE COLLECTIVE CONTENT OF MY MATH-A-NATION course taught to gifted children, grades 4 to 8.

The popular course has been offered through the Super Saturday Program [see SuperSaturday.org] for many years on a campus of the University of Cincinnati. The Super Saturday Program has been offered by PAGE, a 501(c)(3) nonprofit for almost thirty years. It is dedicated to the special educational needs of intellectually gifted children, ages 4 to 14. The students and their parents have grown to love the Math-A-Nation course and have helped add to and polish its content over the years. Parent and student feedback always mentions that these math topics and techniques are not available to them through traditional educational channels.

Overall, I have written this book series to insure that my children, and yours, are equipped with these important math concepts for mental training designed to insure their preparation in the modern competitive world. Parents tell me that this "math thinking" helps their children acquire a mental edge. In turn, this mental edge, allows them to excel in their math and science studies, and also prepares them for advance problem solving in the real world.

The Super Saturday brochure describes this course in this way:

> "With a little imagination and lots of fun you can take on the world of everyday math like never before. How quickly can you prove that 54,879,654 + 52,418,793 = 107, 298, 447? How about instantly shouting out the square of 35? Best of all, you can do it without having to create a math problem – just yell the answer out! If math scares you, fear it no more. If math excites you, come along and learn even more. You will be entertained as you learn new ways to explore math. *Prerequisite*: Must be able to add and multiply single digits from one through nine."

The methods discussed in this book can be used by anyone interested in developing their mental math skills. Also included is a lot of fun and knowledge about the history, people, and development of mathematics.

I need to recognize several people who have been very instrumental in the development of this book.

- Dave Cantey, former CEO of Super Saturday, who asked me to create and teach challenging courses in mathematics for gifted children.
- Saundra Mitchell for her early editing efforts.
- Richard Irwin, my oldest son, for his verifying all math and edits.
- Janet and Brian Hurn who helped design the book cover.
- My family for their patience during the months I dedicated to this book.
- Past students and parents who have, through suggestions, made this a better book.

A final thought. The book title, "*Math 4 2-Day*," was developed to offer the reader a challenge. Obviously, it is a play on words, using numbers, 4 and 2 to represent the words <u>for</u> and <u>To</u>day. Yet it is so much more.

Look closely at the number, 42 (forty-two)! Yes, it is an integer, or natural number, that lies between 41 and 43. 42 is a pronic number; that is, a number that is a product of two consecutive numbers $(x)(x+1)$, 6 x 7 = 42. It is a sphenic number; a positive integer made up of the product of three distinct prime numbers (2, 3, and 7; 2 x 3 x 7 = 42). It is also an abundant or excessive number. (I'll let you look this one up!)

42 is also a perfect score on both the USA Math Olympiad (USAMO link: http://www.usamo.org/) and the International Mathematical Olympiad (IMO link: http://www.imo-official.org/).

It is the angle of degrees for which a rainbow appears.

42 is also found several times in the Christian Bible – 42 is the number of generations (names) in the Gospel of Matthew, describing the genealogy of Jesus. Revelations prophesied that the Beast will dominate over the Earth for 42 months.

Lewis Carroll, of *Alice's Adventures in Wonderland* fame, used 42 over and over in his novels. He was also a mathematician and is known as Charles Dodgson in the mathematical world.

So the number 42 is quite interesting. It, like all numbers, is unique. It has even more exciting properties that you can explore on your own.

In closing, the number 42 is found in the novel *The Hitchhiker's Guide to the Galaxy*, by Douglas Adams. 42 is the answer to the "Ultimate Question of Life, the Universe, and Everything" as calculated by a special supercomputer over 7.5 million years.

Have ideas or just a comment you would like to share with the author? You can eMail him at: **mrirwinreg@usa.net**

Introduction

M ATH 4 2-DAY IS A JOURNEY! It is a collective work of a course, titled Math-A-Nation, that is taught over six Saturdays to intellectually gifted children, grades 4 to 8.

The book does have a structure. You can read it from front to back, Day One through Day Six, or jump around reading those topics of interest to you. Think MATRIX. That is, the Day numbers are like columns in a matrix, while the major sections and topics are the rows. Thinking about it this way allows you to move around in the matrix any way you want. Explore! Keep in mind that the matrix tends to move from top to bottom and from left to right. Left to right means that a topic begun in the Day (column) will continue its development through successive columns (Days).

Take a moment to look through the Table of Contents to familiarize yourself with the structure and decide how you want to attack the book. Again, Explore! It's fun.

What does this mean for you?

Well, as every good student already knows, the most important, and probably the biggest muscle, is his/her BRAIN. This book, and the material contained within, is designed to stretch, massage, and exercise that great brain muscle of yours. "Math Thinking" is developed by doing math, just like any other muscle develops through continual exercise. Improved "math thinking" allows you to become a better problem solver. So what? Have you looked outside lately? The world desperately needs better problem solvers. This world needs you to develop your brain muscle in order for you to step-up and solve its problems.

So, think of this book as an adventure through mathematics. Just use your imagination as you take this journey.

Let's get started on the journey. Hup ... Two ... Three ... Get ready to exercise your brain muscle.

Acknowledgements/Credits

SOME OF THE PHOTOGRAPHS AND LINE DRAWINGS INCLUDED IN THIS BOOK ARE THE WORK OF OTHERS! It is my intent to acknowledge their kindness by allowing my inclusion of their work in this book.

Follows are the photographs and line drawings that are the copyrighted work of others:

- Get Off the Earth Puzzle (pages 24 and 25). This puzzle was created by Sam Loyd. Sam Loyd, Get off the Earth and puzzle design are registered trademarks of The Sam Loyd Company and used by permission. The official website is samloyd.com.

- CUPID's ARROW (pages 119 and 120). This puzzle was developed by Dr. Clifford Pickover and can be found in his book, *Wonders of Numbers*, published by Oxford University Press, USA (June 2002). This puzzle is used by permission. His official website is pickover.com.

- Jill Britton, teacher, consultant, author, and conference speaker, on topics of mathematics. She has granted permission for use of several photographs from her website, titled, Golden Sections in Art and Architecture; located at britton.disted.camosun.bc.ca/goldslide/jbgoldslide.htm. The photographs are:

    Mona Lisa (page 135) – note rectangles added by author
    Mond Crucifixion (page 136) – note triangle and pentagram added by author
    Holy Family (page 136) – note pentagram added by author
    Bathers (page 137) – golden subdivision and rectangle added by author
    Parthenon, Acropolis, Athens with rectangles (page 137)

- Klein bottles© (page 175) Both Klein bottle pictures are the work of Cliff Stoll of ACME Klein Bottles in Oakland, California. They are used here by permission. You can learn more about these bottles at his website: kleinbottles.com.

Day One

W ELCOME TO MATH 4 2-DAY! Today, the first day of the course, we will be setting the tone for the class and, of course, a brief review of things you already know.

Each new day will offer a limited introduction in the form of an *Outline*. The outline will show what will be covered during that day. In addition, it will include the *Objectives* for this unit. Here are the Outline and Objectives for Day One.

Today's Outline

Quotations of the Day
Bit of Fun in Math
> 10 Commandments of Math

Review of Types of Numbers
> Zero's special role - importance

Avoiding Careless Error Math (Part 1)
Simple Math is All You Need (Part 1)
> Fun with numbers – 10 / 5 / 2, x 11; AB x YZ

Calculator Tricks – (Part 1)
Bits of Algebra – The beginning
Endgame for the Day
> Get off the Earth puzzle!

Objectives:

(1) Build self-confidence with math
(2) Build / enhance skills to avoid careless errors
(3) Re-enforce knowledge about the 'laws & rules of math'
(4) Increase personal skills using 'brain math'
(5) Make math 'feel' equal to playing a game
(6) Build a lifetime passion for math

Quotations of the Day

ċ

I hear and I forget.
I see and I remember.
I do and I understand!

-- *Chinese Proverb*

ċ

Numbers are the highest degree of knowledge.
It is knowledge itself!

-- *Plato*

Plato (428/427 BC– 348/347 BC) was a Greek philosopher who was instrumental in founding modern day Western philosophy. Plato, with his mentor, Socrates, and his student, Aristotle, collectively shaped modern philosophy. As a famous mathematician, he founded the first institute of higher education – the Academy of Athens.

ċ

The essence of mathematics is not
to make simple things complicated,
but to make complicated things simple.

-- *S. Gudder*

Stan Gudder (present day) is an American mathematician and computer scientist, from the *University of Denver, Department of Mathematics*. He is known for his research in sequential measurements, Fourrier transformations, Quantum Entropy and Mechanics. Additionally, he is well known for his work in Quantum computing and duality quantum computing.

ċ

Bit of Fun in Math

This section will have at least one topic each day that should interject some levity. Hopefully you will look forward to reading this section each day of the course.

10 Commandments (or concepts) of Math

Teachers, comedians and others formulate their own "10 Commandments" or general guidelines, for a given topic. Here are mine for Mathematics (I've even thrown in an extra):

The Big Ten of Mathematics

1. YOU must read *the complete problem* every time!
2. YOU must remember your math from last year.
3. YOU must know ALL the Rules of Zero (0).
4. YOU must use "COMMON SENSE" or your fathers will become younger than your sons!
5. YOU must master each step BEFORE jumping to the next.
6. YOU must do to one side of the equation, what you do to the other! (balance is key)
7. YOU must remember to always expand –factor/ simplify –combine – and THEN solve the problem.
8. YOU must copy the problem fully and correctly; before declaring that the book is wrong!
9. YOU must check your work again and again to avoid simple careless errors.
10. If YOU do not know how, you shall look it up!   AND

11. If you can't find it or solve it – you WILL ask the *All-Knowing* teacher!

If you follow these 10 Commands/Concepts (plus one) you will always succeed in performing mathematical calculations.

7 times 13 always equals 28!

Two comedians, Lou Costello and Bud Abbot did a math routine where Lou proved to Bud that 7 times 13 ALWAYS equals 28. You can see this 'proof' and method in the 1941 film *In the Navy*. Lou Costello shows Bud Abbot his method of multiplication using two different ways to prove it.

Here is the logic behind the method (according to Costello):

```
  13
 X 7          7 times 3 is 21
  21          7 times 1 is  7
 + 7          21 plus 7 equals 28
  28
```

To prove that his multiplication was correct he then worked the problem backwards –

```
    013         We prove it by dividing 7 into 28
 7 / 28         7 doesn't go into 2 (put 0 above it)
   - 7             divide 7 into 8, get 1 (put above the 8)
    21          NOW subtract 7 from 28 (put 21 below it)
  - 21          DIVIDE 21 by 7 = 3 (put above top line)
     0          You have nothing left - DONE!
```

Finally he proved it is correct by using simple addition:

```
   13
   13
   13
   13        (3+3+3+3+3+3+3 = 21)        21
   13        (1+1+1+1+1+1+1 =  7)       + 7
   13                                    28
 + 13
   28
```

So there you have it –

7 times 13 always equals 28!

Right?

NO! It's all in the use of Zero! (more later)

Review of Types of Numbers

Before delving into the fun ways of learning mathematics, we need to understand what numbers and mathematics are – more precisely, the many types of numbers.

What are Numbers?

Just for the sake of argument, let me state that numbers and mathematics are figments of our imagination; they don't really exist! Can you agree with this statement? Most of you are shaking your heads and saying, "This guy is nuts – numbers are real and mathematics is real – so they exist. My teachers told me so!"

Well I can't argue with that logic. However, let me start by saying, if you think about it, it can be shown that numbers, as we use them, are based on nothingness.

If you look at an empty table and ask someone something like, "How many books are on the table?" They will likely answer, "0" or "none"; and they would be correct. If you ask them to write down how many are on the table, they will write down "0". But what is 0? First, zero (0) represents nothingness. Of course 0 also has a special role in the world of mathematics as a place holder. [More later!] The other nine digits – 1 through 9, represent some value or quantity from one to nine; remember - zero represents the absence of a value or quantity.

Are there other ways to represent numbers? Of course there are – one way is by using our fingers. For instance, a quantity like four can be shown by raising four fingers. In addition, we could create our own symbols to represent each number (each symbol would have to be different for every number). We would grow tired of creating symbols and the list of symbols would be too large for us to remember. (Imagine trying to multiply numbers - you'd have to remember ALL the symbols for every number and memorize an infinitely long multiplication table.)

> *Note*:
> The system of numbers used today was brought to Europe in 1202, when the mathematician Fibonacci introduced the current numbering system to Rome. His book *"Liber Abaci"* introduced Arabic numerals, the use of zero, and the decimal place system to the Latin world. What about before Fibonacci? Before that, there were many different numbering systems used around the world. A good example are the symbols (I, V, X, L, C) used by the Romans. Some of the systems being used around the world were based on 10 symbols; some, like the Mayan system –used 20, and some used as many as 60 symbols!

Eventually everyone agreed to use the ten symbols, 0 through 9, the concept of place-holder notation (later in this section), and use these symbols in combination to represent specific values or quantities when performing math calculations. It was universally decided to use this system of ten symbols to represent all the numbers we write down and use.

In summary:

- Think of numbers as the letters of math (and guess what; there are only 9 of them, oh and the special symbol 0.)
- The ten symbols used today can be combined to collectively represent numbers with an infinite range.

Numbers (many different Sorts and Types)

Ever wonder how integers, whole numbers, rational numbers, transcendental numbers and counting numbers relate to each other? Answer: They build one upon the other! Here is a simplistic explanation of each number type we work with everyday.

Natural Numbers (or the Counting Numbers)

After number symbols were universally defined and agreed to (0, 1, 2, ... , 9) the world of numbers opened up. *Natural numbers* are one type of numbers defined. These are the most fundamental of all numbers and are also known as counting numbers. You may also hear the term Cardinal number when speaking of natural numbers.

> *Note*: *Cardinal*, *Ordinal*, and *Nominal*
> Cardinal numbers tell "how many" (natural or counting numbers). Ordinal numbers tell the order of something (2nd place). Nominal numbers name something – zip code or SSN.

They start with 1, 2, 3, and go on forever; 0 is *NOT* a natural number.

Using these numbers, the basic notion of infinity (numbers that go on forever) is possible. It is because of infinity that the set of natural numbers is essential to mathematics.

Is it possible to count all natural numbers? No! Mathematicians have given a name to this "infinitely" large number that represents how many natural numbers there are. It is called "*aleph nought*" (aleph is the Hebrew letter "A" nought represents "null"). It is a different number from what we are used to; because of its definition. Mathematicians use it to define the bounds of natural numbers. This lets people reference various degrees of infinity.

Whole Numbers

Whole numbers are easy, they start with 0 and go on to infinity. They are all counting numbers and zero. These are the numbers we use in everyday life. We call our numbering system – *base 10*. (Base 10 will be covered on another day.)

<u>Note</u>: *The Number Zero*

The number zero has an interesting history. The ancient Mayans are believed to be the first to introduce the concept of zero as a number. Why did we have to introduce a symbol to denote "nothing" or a "zero quantity" of something? The answer is obvious today – to represent a value of no-value and to be able to separate negative numbers from positive. Of course, it is also used for resetting valuation in columns when combining digits together to represent larger numbers.

Zero's Special Role – importance

At this point we should agree that the symbol zero (0) represents a lack of a value or quantity. For example, there are zero, or no, apples in the basket. We should understand that there is an infinite quantity of numbers; yet there are only nine value symbols and zero to represent those numbers.

How is this possible? It is possible due to the symbol 0. Zero (0) can also be used as a positional holder – used in a very clever way to represent numbers beyond the initial nine digits. It lets us create larger value numbers. It is called *place value notation*.

Place Value Notation

When we count or display numbers above 9, we use a process known as *place value notation*.

Place value notation specifies a hierarchical method to use when displaying (writing down) numbers. It starts from the right digit and moves left. To understand place value notation, simply consider the following which shows how we write numbers which require multiple columns:

9   10   11   - - -   99   100

Using the above list of numbers, we read from left to right – 9, 10, 11, ... 99, 100. A number is read from left to right; yet individual valuation for each symbol (digit) of the number is ranked from right to left. As the above demonstrates, the rightmost position represents the units, to its left is tens, and left of the tens are the hundreds, and so on.

So when using place value notation, you assign a value to each symbol according to where it is in the number (its place). Although the lowest value is the rightmost digit, we read a number left to right.

Let's use an actual example:

Example: 1234   (base 10)

In the example, the symbol 4 (rightmost) is worth 4 units, BUT 3 is worth 3*10 units (or tens), the 2 is worth 2*100 units (or hundreds), and the 1 is worth 1*1000 units (or thousands).

This process is place holder notation. It is brilliant; yet simple. The positions of individual symbols in a number represent different values (units, tens, hundreds and so on.) Using this notation allows us to represent any number (small or large) with a very short list of ten symbols (0, 1, 2, 3, 4, 5, 6, 7, 8, and 9).

> *Note*:
> Remember, in order to represent the entire set of whole numbers in place value notation you NEED the number zero. By using zero we can specify a non-value in a specific position of a number. For instance, you can say that you have NO units of 100 in the number 1025. In fact you wouldn't even say zero hundreds when you announce the number: One thousand twenty-five. (Notice you didn't say – One thousand, no hundreds twenty-five; since there are no hundreds you don't say the value in the number.)

This completes our representation of natural (counting) and whole numbers very nicely *and* by using place holder notation we have many new numbers to work with.

However, this is only the beginning of number types that exist in mathematics.

Integers

To get the *integers*, just add all the negative numbers to your list of whole numbers! Integers include the entire set of negative numbers (before 0), 0, and all positive (beyond 0) numbers:

$$... -3, -2, -1, 0, 1, 2, 3, ...$$

> *Note*:
> The set of natural numbers are often called positive integers; and whole numbers are called non-negative integers.

Up to this point, numbers have been restricted to integers (negative and positive numbers, including 0). You can define many other different kinds of numbers as well!

Rational Numbers

The next series of numbers that can be defined are known as *rational numbers*. They are defined as numbers that consist of the *ratios of integers*; thus the name *rational*. 1/2 is a rational number. 2/3 is also a rational number. Note that all integers are rational numbers, because you can think of them as the ratio of themselves to 1, as in 2 = 2/1 which is certainly the ratio of two integers, and so 2 is a rational number as is 63 (63/1).

How about fractions? Are they rational? Technically, fractions (from the Latin *'fractus'* - broken) are numbers that represent part of a whole. In mathematics, the set of all fractions is called the set of rational numbers.

So yes, fractions are rational!

Note:

The story of rational numbers is interesting. Ancient Greek mathematicians, known as Pythagoreans, were very fond of rational numbers. In fact, when they discovered other numbers which were not rational, they swore that "terrible" discovery to secrecy. If anyone, within the group, spoke of any non rational number to another, they would be executed. Interestingly, shortly after swearing everyone to secrecy, one of the members was discovered, dead, near the river bed.

Irrational Numbers

The square root of 2 is an *irrational number*. It can't be written as the ratio of two integers. Here's the square root of 2 to the first 25 decimal places:

$$\sqrt{2} = 1.4142135623730950488016887 \ldots$$

Transcendental and Algebraic Numbers

Although *transcendental* and *algebraic numbers* are not a part of this course they need to be mentioned since they are types of numbers found in the mathematical world.

To understand *transcendental* numbers, you need to first understand another type of number called *algebraic* numbers. A number is called algebraic if it is the root of a polynomial (of any degree) with rational coefficients. Any number that is not algebraic is called transcendental.

2 and 3/2 are algebraic (they are also rational) because they are the roots of the rational polynomial $3x^2 - 8x + 4$ (in mathematical terms, the polynomial is rational because the coefficients are rational numbers.)

$$3x^2 - 8x + 4 = 0 \text{ is the same as } (3x - 2)(x-2) = 0 \quad \text{so} \quad x = 2 \text{ or } 2/3$$

Observe that the square root of 2 is algebraic (it is also irrational) because it is a solution of the rational polynomial.

$$x^2 - 2 = 0$$

In contrast, a transcendental number is an irrational number that is *not* algebraic. A classic examples of transcendental numbers (that is, not algebraic) are *Euler's* constant (e) and the number *pi*. Both numbers are shown below (the first 25 decimal digits of accuracy.)

Remember: both are irrational (you can't write either as the ratio of two integers.)

$$e \approx 2.7182818284590452353602875 \ldots$$
$$pi \approx 3.1415926535897932384626433 \ldots$$

This sums up the review of numbers and the different types of numbers.

9

Avoiding Careless Error Math (Part 1)

This is the first of three parts that describes a method that can help each person avoid those everyday math errors in addition, subtraction, and multiplication.

Definitions

DIGIT –

A digit is any number 0 through 9. These ten digits are a subset of whole numbers. They are the "single symbols" used to write all other numbers.

ZERO –

Zero is unique, from other digits, because it has a different meaning than the others. The digits 1 through 9 are 'quantity' digits – they represent a specific quantity or value. Zero represents either nothing and/or serves as a *'potential quantity'* indicator. This means that it *may represent* a quantity depending upon its position in the number. For instance, 10 is the next number after 9. This means that zero also serves as a "place keeper." It is used to maintain a column position in the number system we use.

ACTUAL DIGIT –

The digits *1 through 9* are "quantity digits;" representing actual values.

POSITIONAL NUMBERS –

We use the Arabic Number system. Individual digits express values based on their position. Each digit in a number represents a position value – unit, 10, 100, and so on. Each higher position is a geometric expansion of the one before it. When a portion's quantity moves beyond the limit of the digits (... 8, 9), the excess 'spills over' to the next higher position and the previous position resets to 0, to start the count again (1, 2 ...). This is known as the 'carry a number' phenomena.

NON- POSITIONAL NUMBERS –

A *non-positional number* is a number than has no positional value. Although it has a face value, it isn't used to represent a quantity beyond its own value. In reality, it is each individual digit within a number. Each *non-positional number* can be added together to get a new number, collapsing them together. This process can be done until you arrive at a single digit. Let me demonstrate:

POSITIONAL NUMBER	Non-Positional Numbers	Final collapsed Number
2,561	$2 + 5 + 6 + 1 = 14$	not yet
. . .	$1 + 4 = 5$	5

Notice that each digit in 2561 is an individual non-positional numbers. Each can be added together until a single, collapsed, non-positional number is obtained.

CHECK DIGIT –

A *Check Digit* is a single digit with a value of 1 through 9. 0 is not a check digit – it has no quantity value. It is obtained by collapsing non-positional numbers.

Understanding Check Digits

The key to avoiding careless math errors is in the use of *check digits*. It is easiest to explain check digits through examples. To create a check digit you take each non-positional number, the individual digits that make up the number, and add them together. If the resultant answer is a single digit, you are done. If the resultant answer is not a single digit, re-do the entire process of collapsing the new number until you have a single digit.

A check digit must be a single digit and it is obtained through summing the individual digits in a number. If the sum of the individual numbers results in more than one digit, the process is repeated.

Here are three numbers and the process of creating a check digit:

A) $35 \rightarrow (8)$ B) 235   C) 123,876

                                    $2 + 3 + 5 \rightarrow 1) \rightarrow (1)$ $6, 21 = \rightarrow 6 + 2 + 1 \rightarrow (9)$

In Example (A) the number is 35 - it has two non-positional numbers (3 and 5). If you add their values together, you get 8. Since 8 is a single digit, you are done. The check digit for 35 is *8*. Example (B) is a more complex; the check digit is 1! If you add $2 + 3 + 5$ the answer is 10. However, 10 is not a single digit – so you add $1 + 0$ together to get a final check digit 1. In example (C) you do a similar process. You can add groups of digits together and then add the sub-groups to determine the final check digit. Adding $1+2+3$ equals a sub-number of 6 and $8+7+6$ equals a sub-number of 21. Then add the 6, 2 and 1 together to get the final check digit of *9*.

> *Note:*
> When working with zeros (0) you simply discard them when adding up individual digits of a number. So, for a number like 5001 you would just add 5 and 1 (dropping the zeros) to get a check digit of 6.

EXERCISE 1

Now you try to come up with the check digits for these numbers. By the way, ODD or EVEN doesn't matter in check digits. Find the check digit for each problem:

22 (4)	42 ( )	51 ( )
74 ( )	65 ( )	19 ( )
102 ( )	437 ( )	746 ( )
292 ( )	122 ( )	992 ( )

- Answers are on the next page -

ANSWERS: (Exercise 1)
22 (4), 42(6), 51(6), 74(2), 65(2), 19(1), 102(3), 437(5), 746(8), 292(4), 122(5), 992(2)

Casting Out Nines

When creating a check digit, if you encounter a "9", or a sum of 9, you can cast it out; like zeros. The reason is because 9 is the highest digit of the nine values (1 – 9) and when you add 9 to any single number, the result will be a two digit number that, when further reduced, will result in the same check digit! For example:

(A) 3 + 9 = 12 (1 + 2 = 3)   (B) 6 + 9 = 15 (1 + 5 = 6)

Every time you come to a 9, drop it.

IMPORTANT: If it is the LAST 9; then KEEP IT!

If after adding all the digits together your last digit is a 9 – keep it. Since a check digit must be between 1 and 9. If the final digit is a 9 – that is your check digit.

Keeping the last 9

Here are three examples of casting out nines (9) and the last one shows the importance of keeping the last (remaining) 9 as a check digit.

(A) 275   (B) 993,097,632   (C) 810,207,540
     (9) + (5) → (5)   (3) + (7) + (2)   (9) + (9) + (9) → (9)
                              (3) + (9) → (3)

Using example (A), when you add the digits, 2+7 = 9, you can disregard the 9 and continue on to the last digit 5 – the check digit! In example (B) you throw out all nines (6+3=9, so throw it also); add the other digits (3, 7, 2 [7+2 = 9 so throw it also]) until you get your final check digit - 3. in example (C) all numbers add up to 9 so keep the last 9!

Remember, Zero (0) is not a *Check Digit*! (If you get a zero as your answer – change it to a 9 since you threw out the last nine.)

Using Check Digits in Addition Problems

Once a check digit is created, you can use it to verify your answer. To use check digits to verify your answer, simply create a check digit for each number in your problem. Then add those check digits together and create a new *"test" comparison* check digit from that answer. This number will be used later to compare to the check digit of the actual addition problem. Finally, create a check digit for your answer. Are the check digits the same? YES – your answer is right!

Addition and Check Digits

Are you ready to use check digits in addition? Of course you are – so let's begin!

Determining a check digit through addition is just a matter of following these steps:

1. Get the Check Digits for each line of your problem and write them alongside each number
2. Determine the *test comparison check digit* by adding the addition number's check digits together and creating a check digit from the resultant answer. If the answer is greater than 10, create a check digit from that resulting number. (Do this FIRST). This is the "test" comparison check answer digit
3. Do the addition of the PROBLEM and write your answer as you normally would
4. Find the check digit for your answer
5. Match the real answer's check digit with the "test" comparison check digit

– If they match (they're the same); the answer is correct! Good for you!

Example:

```
PROBLEM          Check-Digit
  1382              ( 5 )
 +987             + ( 6 )
  2369┐             ( 2 ) ◄── Test comparison check digit
      └► Answer Check digit = ( 2 )
```

NOTE: (2) equals (2)

IT AGREES … SO THE ANSWER IS CORRECT !

Now you are ready to try a few for yourself.

EXERCISE 2

38	(2)		47	( )	
+ 14	+(5)		+ 14	+( )	
52	(7)		( )		
(7)		They Match	( )		Match?

347	( )		468	( )	
127	( )		421	( )	
104	( )		34	( )	
+241	+( )		+ 7	+( )	
	( )			( )	
( )		Match?	( )		Match?

- If the two numbers do not match; re-do the problem-

Simple Math is All You Need (Part 1)

This section will offer general tips and tricks you can do with day-to-day math. It is the author's intent to get you thinking mathematically all the time. You should try playing games with certain math concepts and challenge your friends to play.

Many of these math tips are intended to help with the rigor of doing math without the use of a calculator. You can expect to see tips that will seem like shortcuts to mathematical calculations. However, there is real math logic behind these "tricks" and you will learn about it as well. Some of the tips will be repetitive from day to day; assuring your understanding and use of them all the time!

This week we will do some simple multiplication; learn a few tricks for multiplying by certain numbers, like 11, and others.

Now let's have some fun with mathematics.

Some Basics of Simple Math

Have you ever played the game of counting up or down by 10, 5, or 2? Of course you have ... let's count up by 2's:

| 2 | 4 | 6 | 8 | 10 | 12 | 14 | and so on.... |

Now try counting down by 2's from 35:

| 35 | 33 | 31 | 29 | 27 | 25 | 23 | and so on ... |

Truly easy stuff to do! How about counting up by 5's starting with, say, 35:

| 35 | 40 | 45 | 50 | 55 | 60 | 65 | and so on ... |

How about making it slightly more difficult and start counting up by 5's from 62:

| 62 | 67 | 72 | 77 | 82 | 87 | 92 | and so on ... |

Not so difficult since the 2 and 7 always repeat – just have to think about it!

And of course 10's starting with any number – perhaps down from 103:

| 103 | 93 | 83 | 73 | 63 | 53 | 43 | and so on ... |

You are probably saying to yourself, I learned these things in my early years of school – either 1st or 2nd grade. That is true. But you probably stopped doing this game around the end of 2nd grade. These mental exercises keep our minds sharp and lead to even better games of multiplication using the same three numbers!

Multiplication Fun with 10, 5, and 2

If you played the above game of adding for a minute, you will have exercised your mind sufficiently to take on multiplying by any of the same numbers quickly. Best of all you can do all the calculating in your mind – *"No calculator need apply!"*

Multiply any counting number by 10

Since this is day one, we will only work with multiplying natural or counting numbers by 10. (Remember they are the positive numbers that have a value.) When you want to multiply any counting number by 10, simply add a zero to the number you want to multiply. So:

10 times 5	50	10 times 32	320
10 times 12	120	10 times 18	180
10 times 154	1540	10 times 2	20

Easy as can be – just add a 0 to the end. How about when multiplying by 100? Well, you just add two zeros to the end of any counting number:

100 times 6	600	100 times 32	3,200
100 times 13	1,300	100 times 18	1,800
100 times 154	15,400	100 times 7	700

To multiply by a thousand (1,000), just add three zeros to any counting number.

Multiply any counting number by 2

Again, we will only work with multiplying counting numbers. To multiply by two, all you have to do is double the value. You should remember your two's table. Remember:

$$2x1=2 \qquad 2x2=4 \qquad 2x3=6 \qquad 2x4=8 \qquad \text{and so on}$$

Using your knowledge of this, with a little practice you can double *any* number – small or large. So here are a few to demonstrate:

$$2 \times 11 = 22 \qquad 2 \times 165 = 330 \qquad 2 \times 1,450 = 2,900 \qquad \text{and so on}$$

Now you try to double these numbers

EXERCISE 3:

2 x 172 = _____   2 x 43 = _____   2 x 521 = _____

2 x 35 = _____   2 x 82 = _____   2 x 27 = _____

- Answers are on the next page -

> ANSWERS (Exercise 3):
> 344 (2x172), 86 (2x43), 1042 (2x521), 70 (2 x 35), 164 (2x 82), 54 (2x 27)

There is nothing to doubling – congratulations you're a doubling expert!

Cutting any counting number in half

Although this section deals with multiplying by a few numbers (10, 5, and 2), we should also cover cutting numbers in half.   Being able to mentally double (multiply by 2) and halve (divide by 2) are powerful tools that will help you every day!

Cutting a number in half is just dividing it by 2 – the opposite of multiplying by 2 (duh!) Before cutting numbers in half, it is good to remember a simple tip –

Tip:
If the last digit (units) of the number is odd (1, 3, 5, 7, or 9) – subtract the last digit by 1 THEN cut that new number in half and FINALLY add a point five (decimal .5) to the answer. The answer will always have a fractional .5 added to the number.

Here are the half values of the first five odd numbers (divided by 2):

Half of 1 is 0.5 (subtract 1, gives 0, half of zero is 0, now add 0.5)
           3 is 1.5 (subtract 1, gives 2, half of 2 is 1, now add 0.5)
           5 is 2.5   (you have the idea ... 7 is 3.5   9 is 4.5)

Here are a few examples of halving odd numbers:

43 / 2 = 21.5   127 / 2 = 63.5   321 / 2 = 160.5

Halving an even number is even easier – just divide it by 2. The answer will always be a counting number! It's just like doubling – only backwards.

Here are a few examples:

32 / 2 = 16   128 / 2 = 64   320 / 2 = 160

Now you try to cut a few numbers in half:

EXERCISE 4 (Half the following numbers – divide by 2)

35 (   )   12 (   )   128 (   )

224 (   )   77 (   )   2231 (   )

- Answers are on the next page -

ANSWERS (Exercise 4):
35 (17.5), 12 (6), 128 (64), 224 (112),77 (38.5), 2231 (1115.5)

Multiplying any counting number by 5

Five is unique– it is the exact middle number of the nine value numbers (1 – 9). It is also exactly half of 10. These two facts make the number five (5) fascinating. Instead of having to learn the entire 5's table, we can use the fact that 5 is half of ten and can quickly solve any multiplication problem involving the number 5.

Remember the easy way to multiply a counting number by 10? Simply add a 0 to the end. To multiply by 5, just take that number and divide by 2 (half). That is the same as multiplying by 5! The trick for multiplying any number by 5 is a two step process:

1. Place a 0 at the end of the number to be multiplied
2. Cut the number in half

That's all there is to it! Sounds too easy? Well it is just that easy!

Here are a few examples to demonstrate the process:

$$32 \times 5 = 320/2 = 160 \qquad 127 \times 5 = 1270/2 = 635 \qquad 32 \times 5 = 320/2 = 160$$
$$7 \times 5 = 70/2 = 35 \qquad 543 \times 5 = 5430/2 = 2715 \qquad 978 \times 5 = 9780/2 = 4890$$

Now it is time for you to do a few. Exercise 5 contains problems for you to solve. Just remember to multiply by 10 (add a 0 to the end) and then cut the intermediate answer in half. That's all there is to multiplying by 5.

EXERCISE 5 (Add a 0 to the numbers; then divide by 2)

$22 \times 5 = $ _____/2 = (   ) $131 \times 5 = $ _____/2 = (   )

$218 \times 5 = $ _____/2 = (   ) $87 \times 5 = $ _____/2 = (   )

$1345 \times 5 = $ _____/2 = (   ) $62 \times 5 = $ _____/2 = (   )

- Answers are on the next page -

Fun with multiplying by 11

Up to this point you have worked with multiplying quickly by 10, 5, and 2. There are a few special tips for multiplying by eleven (11) as well. Today you will learn two new tricks to help you quickly multiply by 11.

ANSWERS (Exercise 5):

110 (22 x 5 = 220/2), 655 (131 x 5 = 1310 /2), 1090 (218 x 5 = 2180/2) 435 (87 x 5 = 870.2)

6725 (1345 x 5 = 13450/2), 310 (62 x 5= 620/2)

Did you know that you can quickly multiply any two digit number by 11? It is so easy you can do it in your head. There is also another method to multiply *any* large number by 11. Doing the second method, you will just write the number down and immediately write the answer beneath it.

Multiply any two digit number by 11

This method is so simple that you will wonder why you never did it before. With a little practice you will multiply 11 times any number (10 to 99) by just writing down the answer!

Here is the two step method:

1. To multiply by 11 just separate the two numbers
2. Put their sum in the middle   . . . DONE!

Of course there are a few rules along the way, here is a step-by-step example:

1. Start with the 2 digit number   54
2. Separate the two digits   5 ____ 4
3. See the hole?
4. Add the two digits together   5 + 4 = 9
5. Fill the hole with the answer   5 _9_ 4

There you have it! ... ... 54 x 11 = 594

<u>Note</u>:

Looks easy, but what happens if the answer, when adding the two numbers in step 4, is greater than 10? If the resultant answer is a two digit number, simply add the "tens" digit to the left number.

59 x 11 = 5 ___ 9   >>> (5+9 = 14)   >>> (5+1) _4_ 9 = 649

73 x 11 = 7 ___ 3   >>> (7+3=10)   >>> (7+1) _0_ 3 = 803

48 x 11 = 4 ___ 8   >>> (4+8=12)   >>> (4+1) _2_ 8 = 528

That's all there is to it!

It is now time for you to do a few problems multiplying any two-digit number times 11.

EXERCISE 6 (Multiplying any 2 digit number by 11)

42 x 11 = _____   83 x 11 = _____

65 x 11 = _____   18 x 11 = _____

77 x 11 = _____   36 x 11 = _____

- Answers are on the next page -

<u>Note</u>:

If you multiply any single digit (1 through 9) by eleven all you have to do is write the single digit down twice! For example 3 x 11 is 33, 5 x 11 is 55, 8 x 11 is 88.

Now that you have mastered multiplying any two-digit number by 11 how about another challenge? Did you know that you can quickly multiply *any* number by 11 with just a little more work? For this, you will probably want to write the number down before you multiply it by 11. Once written, you can just write the answer immediately below it.

Multiplying any number by 11

To multiply any number by 11 you would do these steps first:

1. Write the number down and then draw a line below it.
2. Add a leading zero (0) to the front of the number
3. Write the answer immediately below it.

To be able to immediately write down the answer you need to follow these steps:

1. Start at the right side of the number (unit's position) and add the neighboring number to its right. If there is no number to the right (as is the case of the units position) you write that number down.
2. Then you move left to the next position (tens) and add the number to the right (units) to its value and write that number down.
3. You repeat this action until you finish the problem.

<u>Note</u>:

If, when adding the two numbers, (the current one and it's neighbor to the right) results in an answer greater than 9, simply add the carry over to the next operation.

Although this seems confusing reading it, an example will make it obvious.

For example, the number 51276 can be multiplied by 11 quickly. Just follow the process on the next page –

ANSWERS (Exercise 6):
462 (42 x 11), 913 (83 x 11), 715 (65 x 11) 198 (18 x 11), 847 (77 x 11), 396 (36 x 11)

51276 x 11 =   051276 <<== place a leading 0 in front of the number
                      6 <<== add 6 to its right neighbor (nothing) drop the answer

                   1
             051276 <<== Add 7 + 6 and put the tens value (1) above the 2
                 36 <<== drop unit value (3) below the 7 (remember the carry)

               1 1
             051276 <<== put tens unit from 1+2+7 (1) above the thousands
                036 <<== drop the unit 0 below the two (remember the carry)

               1 1
             051276 <<== add 1 (carry over) +1+2 1 and drop the answer
               4036 <<== drop the 4 below the one

               1 1
             051276 <<== nothing above the 5 so add 5+1 and drop the answer
              64036 <<== drop the 5 below the five

               1 1
             051276 <<== nothing above the 0 so add 0+5 and drop the answer
             564036 <<== drop the 5 below the zero

And the answer is 564036 ...(did you hear the drum roll?)

Another example is in order to clarify the process. Take the number 5345 x 11 and the answer is:

             05345 <<== draw the line and add a leading 0 then work it out
             58795 <<== until you get the answer!

Did you notice that you had no carryovers to deal with in the second problem? Hopefully you have the idea of how to multiply any number by 11. Simply write the number down; add a leading zero; then, starting at the right digit – add each digit to the one next to it on the right; finally write the solution down. Remember: If the number created by adding neighbors is greater than 9, carry the one from the tens to the next sequence of addition.

Now it is your turn to do a few problems to ensure your understanding.

EXERCISE 7 (Multiplying any number by 11)

$$5265 = \qquad \overset{1}{\underline{05265}} \qquad\qquad 7814 = \quad \underline{\hspace{3cm}}$$
$$57915$$

$$1293 = \quad \underline{\hspace{3cm}} \qquad\qquad 13652 = \quad \underline{\hspace{3cm}}$$

$$730 = \quad \underline{\hspace{3cm}} \qquad\qquad 294 = \quad \underline{\hspace{3cm}}$$

- Answers are on the next page -

<u>Note</u>:

Did you know that there is another way to multiply by 11? You can quickly multiply any number by eleven by simply adding a zero to the end of the number and then adding the original number to that number. For example: 324 x 11 = 3240 + 324 = 3564. Another example would be multiplying 4652 x 11 = 46520 + 4652 = 51172. This is simply multiplying by 10 then by 1 and adding the results together. (11 = 10 + 1)

Multiplying any two digit number by another two digit number

The tip for this section is a quick way to multiply any two digit number by another two digit number; that is, multiply any two numbers from 10 x 10 through 99 x 99.

To multiply any two numbers (56 x 74) follow these five steps:

1. Multiply the units values together (6 x 4) and write down the UNITS part of that answer at the extreme right (remember the tens value) ... ___ 4 (remember the 2!)
2. Multiply the tens values together (5 x 7) and place the whole VALUE to the left of the units number – placing a space in between them. ... 35__ 4
3. Now cross multiply the units from one number with the tens of another and then the tens with the units from the other. (5 x 4 = 20 and 7 x 6 = 42). Add these values together (20 + 42) and then add the remainder (if any) from the units calculation. ... (20 + 42 + 2) = 64.
4. Write the units value from this calculation in the middle and add the tens unit to the value for the multiplication value of the tens in step 2 ... (35+6) <u>4</u> 4   = final answer of 4144

ANSWERS (Exercise 7):
85954 (7814 x 11), 14223 (1293 x 11), 150172 (13652 x 11), 8030 (730 x 11), 3234 (294 x 11)

Two problems to demonstrate the process – 92 x 48 and 34 x 76

	92	76
	x 48	x 34
Step 1:	_____6 (remember 1)	_____4 (remember 2)
Step 2:	36__6 (still have 1)	21_4 (still have 2)
Step 3:	36__6	21_4
	(4x2 [8]+ 9x8 [72]+1 = 81)	(3x6 [18]+ 7x4[28] + 2 = 48)
Step 4:	(36+8)_1_6	(21+4)_8_4
ANSWER:	4416	2584

That's all there is to it – multiply the units – place the unit part of the answer (minus carryover) to the extreme right; then multiply the tens – place that answer to the left of the units, leaving a hole in between; and finally, cross multiply, dropping the sums into the middle. Add all carryovers as necessary.

Multiplying by powers of 2 – 2, 4, 8

Since you have gotten this far, a bonus tip is deserved.

Did you know there is a rapid method to quickly multiply any number by 4 or 8? All you have to remember is that 2 x 2 = 4 and 2 x 4 = 8; so 2 x 2 x 2 = 8.

Using this knowledge and knowing how to double any number, you can quickly multiply any number by 4 or 8. Have you guessed how? Just double a number twice to multiply by 4. As for 8, just double three times! Here are a few examples: Remember to multiply any number by 4 just double twice and double three times to multiply by 8.

4 x 16 = (1st double) 2 x 16 >> (2nd) 2 x 32 = 64 is same as 4 x 16 = 64

4 x 23 = (1st double) 2 x 23 >> (2nd) 2 x 46 = 92 is same as 4 x 23 = 64

8 x 12 = (1st) 2 x 12 >> (2nd) 2 x 24 >> (3rd) 2 x 48 = 96 is same as 8 x 12 = 96

8 x 33 = (1st) 2 x 33 >> (2nd) 2 x 66 >> (3rd) 2 x 132 = 264 is same as 8 x 33 = 264

It really is this easy. You just have to remember that 4 times anything is the same as doubling twice and 8 times anything is just doubling three times!

Calculator Tricks (Part 1)

The calculator is a great tool for performing calculations. It can also be used to do some interesting numerical tricks. Here are a few. Just follow the steps in each problem.

A word of warning – make sure you press the (=) sign after every operation!

Is that your final answer?

1. Have someone pick a number between 1 and 9
2. Now have them use a calculator to first multiply it by 9, and then multiply it by 12,345,679 (notice there is no 8 in that number!)

Have the person show you the result so you can tell him the original number he selected! How? Easy!

Answer: Display will show nine identical digits of their original number; e.g. 333,333,333 will be 3 and so on.

This trick is based on the fact that 9 x 12345679 = 111111111. When you do it, you are actually multiplying your original digit by 111111111. By the way, that *8-digit number (12,345,679)* is easily memorized: the 8 is missing and there are 8 digits.

The Birthday display

1. Multiply 4 times your birthday month - (press =)
2. Add 13 - (press =)
3. Multiply by 25 - (press =)
4. Subtract 200 - (press =)
5. Plus Day of Birthday - (press =)
6. Multiply by 2 - (press =)
7. Subtract 40 - (press =)
8. Multiply by 50 - (press =)
9. Add last 2 digits of birth year - (press =)
10. Subtract 10500 - (press =)

Answer: Displays birthday in form MM or M then DD YY format

Cheat the Calculator

This trick may not work. It works best with cheap eight display calculator! Steps:

1. Pick a number – enter it into the calculator
2. Press multiply (ONCE) then immediately press divide (ONCE)
3. Press the equal (ONCE) = sign until the answer gets to zero   S T O P ! -
4. Press the multiply (ONCE) then immediately press plus (ONCE)
5. Press the equal (ONCE) = sign

Answer: Displays original number put in calculator

Bits and Pieces of Algebra

Algebra scares many people; but it doesn't have to. Using Algebra we can do a few neat number tricks. Here are two for you; you can use a calculator if you want!

The magic number 5 (also known as ALL THE SAME)

STEPS	EXAMPLE
1. Think of any whole number (0, 1, 2, and so on)	7
2. Add the next higher number in sequence to it	(7 + 8) = 15
3. Add 9 to the answer	(15 + 9) = 24
4. Divide the answer by 2	(24 / 2) = 12
5. Subtract the original number	(12 − 7) = 5

The answer is always 5! Algebra shows us that this is true.

If you do the same steps with the value x as an unknown to represent any number you will realize that each step can be shown as an expression –

Step 1: is x; Step 2: is $x + (x+1) = 2x+1$; Step 3: is $2x + 1 + 9 = 2x+10 = 2(x+5)$;
Step 4: is: $2(x+5)/2 = (x+5)$; Step 5: is $x+5 − x = 5$!

Here is another problem that is similar:

Take any number and (1) Add 7; (2) Multiply by 2; (3) Add 4; (4) Divide by 2; (5) Subtract the original number that you started with.

The answer will always be? _____

Did you get 9? If not think about the steps with the variable x – (1) $x+7$ (2); $2(x+7) = 2x + 14$; (3) $2x+14+4 = 2x+18 = 2(x+9)$; (4) $2(x+9)/2 = x + 9$; (5) $x+9-x = 9$. So it will always give you 9!

Reading your mind

This algebra trick has the person choose a number, do a few calculations and then give you the answer. With the answer, you do some simple math and give the original number.

Here is the trick and the mathematics behind it.

STEPS	EXAMPLE	Algebra Example
1. Choose any whole number	8	x
2. Double the number	16	$2x$
3. Add 4 to it	20	$2x+4 = 2(x+2)$
4. Divide by 2	10	$2(x+2)/2 = x+2$
5. Add 13 to it	23	$x + 2 + 13 = x + 15$

Ask them the number they have after doing these five steps. Then subtract 15 and tell them their original number!

Endgame for the Day

This section will often have a fun puzzle or activity that you can do. It will usually involve mathematics.

Today's activity is working with a puzzle that has been around since 1896.

Get off the Earth puzzle!

This puzzle is the first of a series of popular puzzles known as vanishing puzzles. Sam Loyd first created, patented (563778) and copyrighted this puzzle in 1896. It has one disk atop another with 13 men (China men) along the circle. By rotating the inner disk, one of the men vanishes – leaving only twelve! This puzzle is considered, by many, his greatest achievement and perhaps the greatest mechanical puzzle invented.

Above is a copy of this famous puzzle. There is a larger copy on the next page that you can photocopy, cut out, and try for yourself. Place the arrow of the inner circle to the N.W. you have 13 men – move the arrow to the N.E. and one man disappears, leaving 12.

This is only one of over 5000 puzzles that Sam Loyd created during his lifetime. All of them can be found in the book *Sam Loyd's Cyclopedia of 5000 Puzzles, Tricks, and Conundrums*. A PDF file can be found at (verified active on 12 May 2011) http://www.mathpuzzle.com/downloads/. You can also purchase a color copy of this and other Sam Loyd puzzles at the Samuel Loyd website: http://samloyd.com/.

Sam Loyd, Get off the Earth, and puzzle design are registered trademarks of the Sam Loyd Company and used by permission.

Day Two

WELCOME TO DAY 2 OF MATH 4 2-DAY! Today, we continue along the same path we began on Day One.

If we were in the classroom, versus reading this book, we would begin with a brief review of the material from day one. But we're not and you can always go back to any part of day one if you need to review. Instead we will push on with today's agenda:

Today's Outline

Quotations of the Day
Bit of Fun in Math
    I have a friend!
    Land of Candy ...
Five Easy Ideas in Math
Avoiding Careless Error Math (Part 2)
Mathematical Curiosities (Part 1)
    Palindromes
Simple Math is All You Need (Part 2)
    Multiplying any number by 9 and others
A few Number Oddities
    The Apocalyptic Number 666
    Is a Billion, a Billion through the World
Calculator Tricks (Part 2)
Math is Logic – or – The Logical Methods of Math (Part 1)
    What is logic and reasoning
Endgame for the Day
    Creating a $1.00 USD Money Shirt!

Objectives:

(1) Continue building skills math to avoid careless math errors
(2) Continue to build on ideas for performing math operations
(3) Build confidence when faced with logic problems
(4) Continue to perform mental math – avoiding the calculator
(5) Relate motor skills with math while building a model

Quotations of the Day

❧ · ❧

Mathematics is the gate and key
to the sciences!

-- *Roger Bacon*

Roger Bacon, (c. 1214 – 1294), known as *Doctor Mirabilis* (Latin: "wonderful teacher"), was one of the most famous Franciscan friars of his time. He was an English philosopher who was credited as one of the earliest European advocates of the modern scientific method; inspired by the works of Muslim scientists!

❧ · ❧

Wherever there is number
there is beauty!

-- *Proclus*

Proclus Lycaeus (Proclus born circa 410 - 412 AD/CE) was one of the last Great Greek Philosophers. He was extremely influential on Christian (Greek and Latin) and Islamic thought. The majority of Proclus' works are commentaries on the dialogues of Plato. Proclus also wrote a very influential commentary on Euclid's first book of *Elements of Geometry*.

❧ · ❧

If a messy desk is the sign of a messy mind,
then what is an empty desk?

-- *Albert Einstein*

Albert Einstein published more than fifty scientific papers and also several non-scientific books. Einstein was a theoretical physicist and in 1999, *Time* magazine named him the "Person of the Century". Einstein received the Nobel Prize in Physics in 1921. Most people know him for his Theory of Relativity ($e = mc^2$).

❧ · ❧

Bit of Fun in Math

Today we have two thought provoking topics. The first is a list of things my friends like and a list of things they do not. It is your job to figure out why. The second is a bit more fun – you try to name the candy bar by the clues given.

I have a friend who ...

Likes	but	Dislikes
Lukewarm		Hot
Chairs		Tables
Backgammon		Chess
Headboards		Beds
Eyewitnesses		Victims
Thumbtacks		Staples
Mistletoe		Holly
Bears		Bulls

Can you figure out why my friend likes certain things and not others?

Land of Candy

Here are a few clues that are associated with ten candy bars. Can you name them?

1. A famous former baseball player _____

2. Twin letters _____

3. All for one and one for all _____

4. Indian Burial Grounds _____

5. Galaxy _____

6. Red Planet _____

7. Sun explosion _____

8. Can't hold on to anything _____

9. A sweet sign of affection _____

10. A favorite day for working people _____

- Answers are on the next page -

ANSWERS: (Bit of Fun, - Day Two)
I Have a Friend Who: The friend likes anything with a body part in the word (arm, hair, and so on)
Land of Candy: (1) Baby Ruth , (2) M & Ms, (3) Three Musketeers, (4) Mounds, (5) Milky Way,
                      (6) Mars, (7) Star Bursts, (8) Butterfingers, (9)Kisses , (10) Pay Day

Five Easy Ideas in Math

There are many ways to perform mental calculations. Look at these four problems and see if you can quickly solve them without using a calculator or pencil and paper:

$$596 + 40 \qquad\qquad 50 \times 17 \times 2$$

$$547 - 138 \qquad\qquad 12\tfrac{1}{4} + 5\tfrac{1}{2}$$

Try to do the problems "in your head." If you solved the problems, think about how your solved them and write down HOW you found the answer for each problem.

Did you know there are a series of basic math methods people use when solving problems? There are five common methods and they are:

- Counting PARTS up and down (adding / subtracting by series)
- Working with compatible numbers first
- Performing equal addition to solve simpler problems
- Breaking apart problems and numbers
- Using compensation

Using these five methods when performing calculations can make doing math "in your head," as simple as: 1 – 2 – 3!

Counting PARTS up and down

Using this method you can quickly and accurately add or subtract numbers. This is especially true when one of the numbers you are working with is a series number (1, 2, 3.) That is, if one of the positional digits is based on a counting value. Specifically, if the units are 1, or 2, or 3, or tens are 10, or 20, or 30, or hundreds are 100, or 200, or 300, and so on. To do the math, just start with the non-series number and count up/down the necessary number of times. Again, an example is in order here to further clarify.

59 + 3	Start with 59 and count UP by 1, three times... 60 (1), 61 (2), 62 (3)
98 – 2	Start with 98 and count DOWN by 1, two times ... 97 (1), 96 (2)
384 + 30	Start with 384 and count UP by 10, three times ... 394, 404, 414
247 – 20	Start with 247 and count DOWN by 10 two times... 237, 227
542 + 200	Start with 542 and count UP by 100 two times... 642, 742

Using the first example, on the previous page, *59 plus 3* you can count up, starting with 59, counting by ones, three times: start at 59 and count – 60 (1), 61 (2), 62 (3). The answer to the problem 59+3 is 62! The fourth example subtracts by tens (2 times); you will count down by tens two times starting at 247 – 237 (-10), 227 (-20). The answer to the problem 247 – 20 is 227!

By practicing this process of counting up or down by 1 (unit, ten, etc.) you can quickly perform most two number addition and subtraction in your head.

You are not limited to a single number when counting up or down in series, like 2, or 30, or 400. You can do this same process with numbers like adding or subtracting 42 or 57. You just need to do the highest value first and then the next lowest value.

The following problems demonstrate mixed column addition/subtraction:

125 + 32	Start with 125 and count UP by (10) three times ... 135, 145, 155
	Now finish with counting UP by (1) two times ... 156, 157
	125 + 32 = 157!
674 – 253	Start with 674 and count DOWN by (100) two times ... 574, 474
	Now count DOWN by (10) five times ... 464, 454, 444, 434, 424
	Now finish by counting DOWN by (1) 3 times ...423, 422, 421
	674 – 253 = 421!

Doing addition and subtraction this way, you perform math calculations from left to right; instead of right to left. It's simply counting up or down by columns. Doing this, you increase by the next highest value as you count (90, 100, and 110). It also forces you to count each position separately (hundreds, tens, units) – opposite of the way you have been taught. It's easy to count by single units (1,2, 3, 10, 20, 30, and so on) – you can even use your fingers to help you solve the problem.

Work with compatible numbers first

Compatible numbers are numbers that you recognize and can quickly work with. Often these numbers will, when added or multiplied together, produce answers that end in 0 – 10s, 100s, 1000s, and so on.

When you see two numbers that you can quickly do mentally, do them first in the problem and then solve for the rest. For example:

47 + 28 + 13	Start with 47 plus 13 (they equal 60) then add 28 to 60 = 88
8 x 7 x 5	Start with 8 times 5 (they equal 40) then multiply 40 x 7 = 280

Performing equal addition to solve subtraction problems

There are times that you need to subtract two numbers that may appear to be difficult at first glance.

For instance:

$$184 - 28 \qquad\qquad\qquad 492 - 185$$

These problems seem challenging; but, they really aren't! You can quickly solve them by using addition. You may ask yourself, "Addition?" YES, addition!

In both examples you can add a number to *each part* of the problem and then solve an easier problem:

$184 - 28$	If you add 2 to each part of the problem you get $186 - 30$, The problem is easier to solve: ... $186 - 30 = 156$
$492 - 185$	If you add 15 to each number, it becomes easy to solve, It becomes $507 - 200$ and the problem is easier to solve ... $= 307$

This works because of a simple rule of math. If you add (or subtract) the same number to each part of the subtraction problem; before solving it, the answer will still be the same. Take the simple subtraction problem, $10 - 2$ and add a value to each number and see how it doesn't affect the final answer in any way. Here is an example, using an unknown (x):

1. Add any value (x) to each part: $(10 + x) - (2 + x) =$???
2. This will expand out to: $10 + x - 2 - x =$
3. This can now be regrouped to: $10 - 2 + x - x =$
4. This finally reduces to $10 - 2$ which equals 8!

Notice in the above steps, any value x, is added to both the 10 and the 2, using grouping to add a value to each. Then the problem is expanded out in step two. Step three then regroups the problem so that the positive x and negative x are at the end of the problem and they will cancel each other out; returning to the original problem of $10 - 2$.

Breaking apart problems and numbers

This tip requires understanding how numbers can be broken apart to do simpler math. This is especially good for multiplication and subtraction.

The following examples show how a problem can be broken apart to make it simpler:

$$56 \times 7 \qquad\qquad 128 \times 3 \qquad\qquad 653 - 328$$

The first problem (56×7) can be broken into two problems and then solved: Instead of multiplying 56×7 first separate it into two problems and add their answers together.

Note:
Remember that 56×7 can be broken into *50×7 plus 6×7*.

Using this tip you can quickly solve the problem. Just follow these steps:

1. Separate the problem into two 50 x 7 and 6 x 7
2. Multiply 50 x 7 = 350
3. Multiply 6 x 7 = 42
4. Add the two answers together, 350 + 42 = 392!

The second problem (*128 x 3*) can be solved by separating the problem into three different problems and adding their resulting answers together. Here are the steps:

1. Separate the problem into three parts 100 x 3, 20 x 3, and 8 x 3
2. Multiply 100 x 3 = 300
3. Multiply 20 x 3 = 60
4. Multiply 8 x 3 = 24
5. Add the three answers together, 300 + 60 + 24 = 384!

The first two problems use multiplication and then sum each partial solution together for the final answer. The next problem is a subtraction problem and a similar process is used to solve it. To solve *653 – 328*, follow these steps:

1. Separate each number into its three parts 600, 50, and 3 (653) – 300, 20, and 8 (328)
2. Subtract the hundreds from both numbers: 600 – 300 equals *300* (intermediate answer)
3. Subtract the tens next: 50 – 20 equals *30*
4. *Add* the 30 from the tens part, to the hundreds: 300 + 30 = 330 (new intermediate answer)
5. Subtract the units next: 3 – 8 equals *NEGATIVE 5*
6. *Add* the – 5 from the units part, to the intermediate number: 330 + (-5) = 325

Notice in step 6 you added a negative number to the intermediate number. In reality if the answer is negative you aren't adding – rather you are subtracting the number from the partial answer.

Using Compensation

The last tip uses a technique known as compensation calculation. If you have a number that can be adjusted to make the problem easier to solve mentally, adjust the number before the problem is solved and then retract the adjustment.

Using an example, *165 – 47*, the process will seem clearer; follow these steps:

1. Add *3* to *47* (ONLY), giving you 50
2. Now subtract 165 – 50 = 115
   You are *not done*! This is an intermediate or partial answer. By adding 3 to 47, you took 3 too many from the problem.
3. Add 3 to the partial answer to solve: 115 + 3 = 118

This example was based on subtracting one number from another and adding a value to one of the numbers to make the problem easier to solve. Because you added a number to only one part of the problem, you needed to correct the intermediary number. In this case, that correction required adding three to get the final answer.

> *Note*:
> You could have used an earlier method, known as "Counting parts up and down," to easily solve this problem. You simply add three to both numbers and then solved the problem: $(165 + 3) - (47 + 3) = 168 - 50 = 118$!

Is the process of using compensation the same for adding two numbers? The concept is; however, the actual steps are different. Try another example, $387 + 58$

1. Add *2* to *58* (ONLY) giving you 60
2. Now add $387 + 60 = 447$
   > You are *not done*! This is an intermediate or partial answer. By adding 2 to 58, you added 2 too many to the problem.
3. Subtract 2 from the partial answer to solve: $447 - 2 = 445$

In this addition problem, one of the numbers was increased by some number. Therefore that number it was increased by had to be subtracted from the intermediate answer. If you decrease one of the numbers in an additional step to easily solve the problem you will need to add that number back in to get the final solution. For instance, take the problem $572 - 353$, you can solve using these steps:

1. Subtract *2* from *572* giving you 570
2. Now, Subtract *3* from *353* giving you 350
3. Now subtract $570 - 350$ getting 220
   > Again, you are *not done*! This is a partial answer. By subtracting 2 from 572 *and* subtracting 3 from 353, you actually took 2 too many from the main number. Then by taking 3 from the subtracting number you added too many to the second number. You need to compensate for these two changes to the problem. So you need to add 2 for the first number *AND* subtract 3 for the change in the subtracting number.
4. Add negative 1 (-1) to the partial answer to solve: $220 - 1 = 219$
   > The intermediate number is adjusted by $+2$ (top number) and -3 (bottom number).

Just remember that when using compensation, if you add a number, subtract it later. If you subtract a number – add it later. Be careful with the final compensation based on this operation.

That concludes these five methods!

Avoiding Careless Error Math (Part 2)

This section is part 2 of 3. It continues the method of using check digits to avoid those everyday math errors in addition, subtraction, and multiplication.

Review: Using Check digits in Addition

Last week you used check digits to verify the accuracy of your addition answer. A check digit is a single collapsed number, formed by adding the individual face values of each digit within a number; resulting in a single value 1 through 9. If the number when collapsing the digits (adding the face values together) is greater than 9 you continued the process of collapsing until you have a single digit.

Understanding this process is critical for continuing on to using check digits with subtraction. Therefore a quick review is in order.

Addition and Check Digits

A check digit was made for each number of the problem. Then a "test comparison" check digit was created from the sum of those check digits. Finally the addition problem was solved and a check digit created for the answer. If the *test comparison* check digit and the *answer* check digit matched, the addition problem was correct!

Here are the actual steps needed to use check digit math with addition:

1. Create a check digit for each number of the addition problem
2. Write their check digits alongside each number
3. Create a "test comparison" check digit by adding the individual check digits together (FIRST)
4. Solve *the actual addition problem*
5. Find the Check Digit for the actual answer
6. Match the answer check digit with the test comparison check digit

Now you should be ready to use check digits with subtraction.
Need to review check digits in addition? Please go back and review the process.

Using Check Digits in Subtraction Problems

Check Digits can also be used to verify that your subtraction problems are correct. It uses a similar process to that used for addition.

In general, to verify the accuracy of your answer, from a subtraction problem, you collapse each number in the problem to a single check digit, then subtract one check digit from the other; finally, after solving the problem, create its check digit and compare it to the test comparison check digit.

This process should sound familiar; it is similar to the process used for addition. The difference is in creating the test comparison check digit. Instead of adding the individual check digits together – you subtract them.

Subtraction and Check Digits

You should be ready to use check digits to verify your answers in subtraction. To work with check digits in subtraction; follow these steps:

1. Determine the Check Digits for each line of the problem and write it alongside each number.
2. Determine the *test comparison check digit* by subtracting the problem's check digits from each other (bottom from top) and creating a check digit from the resultant answer. (FIRST)

   <u>Note</u>:
   If the lower number is greater than the top number, you will need to add nine to the top number – before doing the subtraction.
   - This will produce a "test" comparison check answer digit.

3. Do the subtraction of the PROBLEM and write your answer as you normally would.
4. Find the Check Digit for your answer.
5. Match the real answer's check digit with the test comparison check digit.

   – If they match; the answer is correct! Good for you!

Example

	PROBLEM		Check-Digit
A	1382	(5) = (5 + 9) ————————→	14
	- 987	− (6) ←— 6 > 5 so add 9 to 5	− 6
	395 ⌐	-- Test comparison check digit ——→	(8)
		↳ Answer Check digit = (8)	

NOTE: (8) equals (8)

<u>IT AGREES ... SO THE ANSWER IS CORRECT !</u>

B	5432		(5)
	− 1101		− (3)
	4331 ⌐	Test comparison check digit ——→	(2)
		↳ Answer Check digit = (2)	

NOTE: (2) equals (2)

IT AGREES ... SO THE ANSWER IS CORRECT !

These two examples demonstrate how to use check digits when performing subtraction problems.

Notice that an extra step was required in example *A* before obtaining the test comparison check digit. In example *A*, the check digit for 1382 is a 5, while it is a 6 for 987. When you subtract the check digits (5 – 6), 6 is greater than 5 and will result in a negative number. Since check digits must have a positive value between 1 and 9, inclusive, you need to fix this problem. To fix it, add 9 to the top number (5) and then subtract 6 from the resultant addition.

The check digit math in example *B* does not have this problem, so just do the subtraction (5 – 3) to obtain the test comparison check digit 2.

Tip:
Instead of adding 9 to the top number when the lower number is larger, you can perform the math (5 – 6 = –1) and then take the difference from the subtraction (– 1) from 9 ... (9 – 1 = 8). You will get the same check digit by subtracting the difference (– 1) as you would by adding 9 to the top number and then subtracting the bottom number from the addition.

You should be ready to try a few for yourself. Try doing these:

EXERCISE 8

38	(2)	182	( )
– 16	–(7) (9 – 5)	– 84	–( )
22	– 5 = (4)		( )
(4)	They Match	( )	Match?

441	( )	468	( )
– 241	–( )	– 87	–( )
	( )		( )
( )	Match?	( )	Match?

- If the two numbers do not match – re-do the problem -

That is all there is to performing check digit math for verifying your answers through addition and subtraction.

Just remember to perform the same operation on your check digits when obtaining your test comparison check digit as you do for the problem. That is, if you are doing an addition problem, add the check digits together to get a test comparison check digit. If you are subtracting, subtract the two check digits from each other to get a test comparison check digit.

You can also use check digit math when performing multiplication. This will be discussed on Day 5, section, "Avoiding Careless Error Math (Part 3)."

Mathematical Curiosities (Part 1)

This section is initially being introduced today, for the first time. It will offer mathematical topics that are interesting yet may not be related to day-to-day mathematical processes.

The topic for today covers palindrome numbers.

The Exciting World of Palindrome Numbers

When most people hear the word "palindrome" they usually think of words and sentences which are spelled the same forward and backward. For example, "*Mom,*" and "*Racecar*"; or, "*Madam, I'm Adam!*" or "*Do geese see god?*" Some are aware of more complicated ones, like, "*Are we not drawn onward, we few, drawn onward to new era?*" Palindromes are also found in the world of mathematics.

A Palindrome number is a number that reads the same left to right (forward) and right to left (backward)! A few palindrome numbers include 272, 54645, 7321237.

Creating a Palindrome from any 3 digit number

Did you know that you can take almost any 3 digit number and add its reverse to itself; *and* by continuing to do this process, eventually create a palindrome number?

Here are the steps to create a palindrome number from a three digit number:

1. Pick a three digit number
2. Reverse its digits and add this value to the original number
3. If this is not a palindrome, go back to step 2 and repeat

For example, take the numbers 341 and 976

341 + 143 = 484	976 + 679 = 1655
484 is a *Palindrome!*	1655 + 5561 = 7216
	7216 + 6127 = 13343
	13343 + 34331 = 47674
	47674 is a *Palindrome!*

Often adding the reversal of a number to itself creates a palindromic number in a few steps. There are some exceptions; one is known as the *196 Problem*.

With the number 196 aside, about 80% of all numbers under 10,000 solve down to a palindrome in 4 or less steps. That percentage is 90% when solving in 7 steps. There are a few rare cases, like the number 89. It takes *24 iterations* before the number 89 becomes a palindrome. In fact the number 89 takes the most steps of any number under 10,000 that has been resolved into a palindrome.

The 196 Algorithm or the 196 Problem

Some numbers never appear to form a palindrome. The first one discovered was 196. The search to resolve this number to a palindrome has been referred to as the *196 Algorithm* or the *196 Problem*. The problem was written about in the 1984 issue of *Scientific American* article, titled the *196 Palindrome Quest*.

On August 12th, 1987, John Walker, founder of Autodesk, Inc. and co-author of Auto-CAD, created a program on a Sun workstation to search for a solution to the 196 Problem. It ran until just before midnight, May 24th, 1990, almost three years, his program ended after performing 2,415,836 iterations on the number 196. These iterations yielded a number that was over a million (1,000,000) digits long – with no palindrome in sight. John made this 1,000,000+ digit number and his program available to people on the Internet, at a webpage titled *"Three Years of Computing"*. Anyone can go to the website and read more about it. It can be found at: (verified active on 12 May 2011)

http://www.fourmilab.ch/documents/threeyears/threeyears.html

Going to the above website anyone can continue the search from his end point.

One person, Tim Irvin, did just that. He used a super-computer and calculated the *196 Problem* to 2,000,000 digits in 1995. On August 22nd, 1995, after about two months of calculations, the program was stopped. Tim's story can be found at his site, titled, *"About Two Months of Computing."*: (verified active on 12 May 2011)

http://www.fourmilab.ch/documents/threeyears/two_months_more.html

By the end of 2001, Jason Douchette took the calculations to 13 million digits. The search continued in August 2004 when Wade van Landingham reached more than 211 million digits without getting a palindrome number.

196 is not the only number, other numbers that do not appear to create a palindrome include – 887, 1675, 7436, 13783, and so on.

Interesting Palindrome facts

Some interesting facts of palindrome numbers include:
- All palindrome numbers with an even number of digits are divisible by 11
- Of the first ten 3 digit palindromes, five of them are prime (101, 131, 151, 181, and 191)
- Of the 90 triple digit palindrome numbers, 15 of them are prime
- Among the 900 5 digit palindromes, 93 are prime
- Among the 9000 7 digit palindromes, 668 are prime

The ratio drops dramatically after that; palindrome primes are in fact fairly rare.
I hope you enjoyed learning about the world of palindromes.

Simple Math is All You Need (Part 2)

Today, we will learn about a great way to multiply any number by 9, 99, 999, or any number of nines.

Multiplying any number of digits by any number of 9's

Like multiplying any number by 11, there is a great and easy way to multiply any number by 9. It is actually a pretty easy method to remember.

In fact, the same method of multiplying by 9 can be used to multiply any number by *any number of 9*s.

Multiplying any number by 9

Before multiplying any number by *any number of 9s*, it is best to learn the rule for multiplying by 9. Here is the process:

1. Append a 0 to the multiplicand (the number being multiplied by 9)
2. Write this number down
3. Subtract the original number from the number you just wrote down

That is all there is to it. Here are two examples to demonstrate the process:

78	< = = One 9 so add one 0 below = = >	283
x 9		x 9
780	< = = 0 at the end of top number = = >	2830
−78	< = = Subtract original number = = >	− 283
702	< = = The *ANSWER* = = >	2547

Using these examples the process is similar. First, drop the top number down and multiply it by 10 – place a zero at the end. Once the new number (number with 0 appended) is written down, subtract the original number from it.

> *Note*:
>
> This method works because any number multiplied by 9 is the same as multiplying any number by 10 then subtracting the number times 1. An example: 55 x 9 is the same as (55 x 10) − (55 x1). Do you remember that multiplication is just a method that replaces the process of addition? If you add 55 to itself nine times, it is the same as multiplying 55 by 9. So adding 55 to itself ten times and then subtracting 55 from the answer is the same as multiplying 55 by 9!

So multiplying any number by 9 is the same as multiplying by 10 and subtracting the number from itself!

Multiplying a number by any series of 9s

Now that you have mastered the process of multiplying any number by nine you are ready to multiply a number by any series of 9s; e.g. 99 or 999 or 9999.

The actual process is the same as multiplying by 9; you add the same number of zeros to the multiplicand (top number) as you have 9s. Once you have written the new number down, just subtract the original number from it.

That is all there is too it! Here are some examples to demonstrate the process:

```
      54              365              5923
    x 99            x 999            x   99
    5400           365000           592300
    - 54           -   365           - 5923
    5346           364635           586377
```

These problems use the same process of adding the same number of zeros to the multiplicand (top number) as there are nines to be multiplied by. To solve the problem you simply subtract the original number from the number with added zeros.

You should be ready to try a few for yourself. Try doing these three in exercise 9:

EXERCISE 9

```
        635                              878
  x     999                        x      99


  _____  .                       _____  .
```

```
       9673                             218
  x       9                        x   9999


  _____  .                       _____  .
```

- Answers are on the next page -

ANSWERS: (Exercise 9)

634,365 (635 x 999), 86,922 (878 x 99), 87,057 (9673 x 9), 2,179,782 (218 x 9999)

Revisit multiplying any number by 2 or any power of 2 (2^2, 2^3, ...)

To multiply any number by 4 or 8 simple double the original number two or three times. Did you know that 2 x 2 (4) is the same as 2^2 or that 2x2x2 (8) is the same as 2^3? Knowing this, you can multiply any number by any power of 2 (for instance 16 = 2^4). All you have to do is double the number the correct number of times. For instance:

55 x 4 = (55 x 2) x 2	OR	55 x 2, then 110 x 2 = 220
125 x 8 = ((125 x 2) x 2) x 2	OR	125 x 2, then 250 x 2 then 500 x 2 = 1000
33 x 16 = (((33 x 2) x 2) x 2) x 2	OR	33 x 2, then 66 x 2 then 132 x 2, then 264 x 2 = 528

So to multiply by 4 simply double it twice; by 8 double it three times; finally by 16 you double the number four times. Remember, 16 is the fourth power of 2, 2^4!

Note:

Since 16 is a two digit number, if you are multiplying another two digit number, you could also use the two digit number by two digit number multiplication tip from Day One.

Revisit multiplying any number by 5

Previously you were shown a two step method for multiplying any number by 5. It was to multiply by 10 first; or adding a zero to the end of the multiplicand (top number); then cut that number in half (dividing by 2).

Hopefully you have played the double and half game since; practicing halving and doubling numbers. If you have, you should be comfortable with halving numbers.

Remember, to halve any odd number (ending in 1, 3, 5, 7, or 9) you drop the number by 1 value, cut it in half and add 0.5 to the answer.

The method for multiplying any number by 5 can actually be reversed. First cut the number in half and then multiply it by 10. Here are a few examples to demonstrate the process:

32 x 5 = 32/2 = 16, now 16 x 10 = 160	127 x 5 = 127/2 = 63.5, now 63.5 x 10 = 635
7 x 5 = 7/2 = 3.5, now 3.5 x 10 = 35	543 x 5 = 543/2 = 221.5, now 271.5 x 10 = 2215

Some people find it easier to halve a number first and then multiply it by 10. The point is there are two methods for multiplying a number by 5. You can either multiply it by 10 (add a zero to the end) and cut in half OR cut in half and then multiply by 10.

Of course there is also the old method of multiplying in the traditional method; multiply each digit by 5 starting from the right until you finish.

A few Number Oddities

There are a few 'odd' number topics! Follows are a few: a little math using Roman Numerals that defies logic; oddities of the number 666; and cyclic numbers.

The Apocalyptic Number 666

When some people hear the number 666 they think *evil*, "the sign of the Beast," anti-Christ, end of the world, the devil! The number is found in occult films and in print to signify ultimate evil. It puts terror in the hearts of many believers.

The origin of its meaning is often referenced via the quotation:

> *"Here is wisdom, Let him that hath understanding count the number of the beast: for it is the number of a man; and his number is Six hundred threescore and six."*
> – (Bible, Revelation 13:17 – 18)

The original Greek form (as found in the Bible) was "$αὐτοῦ\ ἑξακόσιοι\ ἑξήκοντα\ ἕξ$" which transliterates to *hexakósioi hexēkonta héx*, meaning "six hundred sixty-six"! But I digress; the point is that the number 666 is often seen as an evil number.

The number 666 can also be obtained through mathematical manipulation of numbers. Some have said that the number is so powerful that the number is divined (having inexplicable perceptive powers) by mathematical manipulation. Here are a few interesting ways to derive the number 666 through mathematical manipulation:

- $666 = 1 + 2 + 3 + 4 + 5 + \ldots + 32 + 33 + 34 + 35 + 36$
  (the total sum of all numbers added together from 1 through 36)
- $666 = 1^6 - 2^6 + 3^6$
- $666 = (6 + 6 + 6) + (6^3 + 6^3 + 6^3)$
- $666 = 1^3 + 2^3 + 3^3 + 4^3 + 5^3 + 6^3 + 5^3 + 4^3 + 3^3 + 2^3 + 1^3$ (A Palindrome)
- $666 = 2^2 + 3^2 + 5^2 + 7^2 + 11^2 + 13^2 + 17^2$

Finally here is one with a twist – needs a calculator to do:

Enter $1800 - 666$ in the calculator and turn it over!   (hEll)

So there you have it – the power of 666 in the world of numbers!

Is a Billion, a billion through the World?

Living in the United States, you have learned that a billion is a one followed by nine zeros – 1,000,000,000.

Have you ever thought about how big a billion really is? Well it is a *really* big number. How long do you think it would take you to count to a billion, saying one number a second? It would take you about 31 years, 8 ¼ months to count to 1,000,000,000.

Note:
To determine how long it would take to count to a billion, at one number a second is relatively simple. Just take the number one billion (1,000,000,000) divide by 60 seconds (60 seconds in a minute). Then divide by 60 minutes (60 minutes in an hour). Take that and divide it by 24 (24 hours in a day). Finally take and divide it by 365.25 (365 ¼ days in a year); then covert the decimal back to months. The formula would look like this:

$$((((1,000,000,000 / 60) / 60) / 24) / 365.25) = 31.69$$

To convert 0.69 years to months you need to convert the ratio of 6.9 to 10 to a 12 month ration or x to 12. Once that is done you can use the formula (6.9 x 12)/10 = 8.28 months!

It will take a big hunk of time out of your life to count to a billion. Imagine counting for 31 plus years with no sleep, no play, no food, no nothing; except counting!

Since this part is about whether a billion is a billion around the world, you may be upset to find out that there are two different billions.

This billion, that will take over 30 years to count, is known as the American billion. There is another billion, the British billion. The British billion originally meant a million-million and still does in England and Germany! At least when you hear people talk about a billion on the street!

It does get a little confusing in the world of mathematics. The reason why is found in the origin of the word billion. The 'bi' prefix in the word billion implies *TWO "million"* written side-by-side, thus:

1,000,000:1,000,000 or   1,000,000 x 1,000,000

This would convert to 1,000,000,000,000 (one followed by twelve zeros). So the British billion is actually, the American trillion! WoW!!

To count to the British billion would take; let me see, hmmm, over 31,688 years. How many generations of people would that take? Too many to count!

Thankfully, the American billion is used through most of the world in scientific and mathematical circles.

Cyclic numbers in the world of mathematics

There are multiple digit integers that when multiplied by any number from 1 to the number of digits, in that number, will always return the same non-positional digits as the original cyclic number.

Are you lost? No problem.

In the next section (*Calculator Tricks – Part 2*) you will perform a trick titled, "Not Quite a Million." The trick uses a "*Cyclic Number*," specifically, 142,857. The number 142,857 is 6 digits long and when multiplied by any digit from 1 to 6 will always result in an answer that contains the same six digits in the answer. The positions of the six digits will change, but the answer will have the same six digits as the number being multiplied – specifically 1, 2, 4, 5, 7 and 8:

$1 \times 142{,}857 = 142{,}857$ and $2 \times 142{,}857 = 285{,}714$ and $3 \times 142{,}857 = 428{,}571$

$4 \times 142{,}857 = 571{,}428$ and $5 \times 142{,}857 = 714{,}285$ and $6 \times 142{,}857 = 857{,}142$

Do you see that every answer has a 1, 2, 4, 5, 7, and 8 in it. Thus a cyclic number!

The ROMAN problem of XI + I = X or 11 + 1 = 10

This is not a trick question. It is possible, using Roman Numbers, to make this problem solve true. Do you know how?

Think about the Roman numbering system: I = 1, V = 5, X = 10, L = 50, C = 100, D = 500, M = 1000, and so on. Using this system, you can also 'short cut' any number near the next by placing a smaller unit symbol before it ... thus IV is 4 and IX represents 9, while XI represents 11. The following chart shows how this works:

Count by	1	2	3	4	5	6	7	8	9
UNITS	I	II	III	IV	V	VI	VII	VII	IX
TENS	X	XX	XXX	XL	L	LX	LXX	LXXX	XC
HUNDREDS	C	CC	CCC	CD	D	DC	DCC	DCCC	CM

So does the chart help? Have you figured it out yet?

Take the problem and put it in the middle of an index card:

$$\text{XI} + \text{I} = \text{X}$$

Now to solve the problem, all you do is turn the card upside down:

$$\text{X} = \text{I} + \text{IX}$$

Is it now correct? Does 10 = 1 + 9?

Calculator Tricks (Part 2)

Here are a few more calculator tricks to try today. Hope you enjoy them!

The Count is IN! *or* It's all about the zeros and a little old decimal

This is two parts and the answer will change based on what you multiply by.

<u>NOTE</u>: You will need a 10 digit or greater display calculator for this to work

1. Multiply 123456789 (1 through 9)
2. Times 80.00000073 and PRESS = (notice one 0 is before decimal, 6 after)

Answer: 9876543210

3. Multiply 123456789 (again 1 through 9)
4. Times 8.000000073 and PRESS = (notice seven 0's after decimal)

Answer: 987654321

Words, Words, Words (up and over)

A Friendly Greeting:

1. Enter 6.2
2. Multiply by 6.2
3. Add 0.23
4. Divide by 50

- Just wanted to say "HI"

hELLO

What's for Breakfast? Which one do you want to eat?

1. Square 10
2. Subtract 40
3. Multiply by 30
4. Add 1033
5. Multiply by 2
6. Subtract 3

EGGS

1. Square 68
2. Subtract 7
3. Multiply by 5
4. Multiply by 25
5. Add 220

ShELLS

Not quite 1,000,000 (start with one less)

1. Enter 9 9 9 9 9 9 (six 9's in calculator)
2. Divide by 7 and call it your magic number
    - 142857
3. Pick a number between 1 and 6
4. Multiply the magic number by number you picked (for instance 4 = 571428)
5. Arrange the numbers from the multiplication from lowest to highest to form a new six digit number (124578)
6. Let me guess – the NEW six digit number is

Answer: 1 2 4 5 7 8

Math is Logic – or – the Logical Methods of Math (Part 1)

Logic is the science of reasoning and correct inference. Inference is the act of reasoning that allows a person to draw a conclusion or make a logical judgment from direct or circumstantial information. Historically, logic has its roots in the teachings of the Greek philosopher Aristotle. During the mid 19th through 20th Centuries it became a major science.

Some argue that logic is a branch of the philosophical sciences, versus math. Logic is an integral part of mathematics, as a subfield (known as symbolic logic) with close connections to computer and philosophical science.

Reasoning out word problems

Logic is really the science of *correct reasoning*. It requires common sense when faced with uncertainty when reading the problem. Here are three interesting word problems that infer some sort of logic:

> I am a nobody ...
>> Nobody is perfect ...
>>> Therefore, I am perfect!

If you read and accept this little ditty as being true, you conclude that you must be perfect! To accept it as truth you must accept the top two lines as valid postulates.

> *Note*:
> A postulate is a proposition (statement) that requires no proof, being self-evident, or that is, for a specific purpose, assumed true. It is a necessary condition or prerequisite for the outcome to be true.

Here is another logic statement that sounds plausible (NOT!):

> Stealing refrigerators is hard work ...
>> Hard work is good for you ...
>>> Therefore; stealing refrigerators is good for you!

So stealing refrigerators must be good for you. Don't worry about those pesky police – just let them know that you are working hard.

Finally, here you need to use a mathematical calculation and think about what it is actually trying to say:

> If it's zero degrees outside today and
>> It's supposed to be twice as cold tomorrow,
>>> How cold is it going to be?

Accepting the math rule: zero multiplied or divided by any number (except 0) is always zero; the next day's temperature will still be zero; *it can't be twice as cold.*

All three of these problems are fun. But are they logical? In other words, do they make sense, common sense? Most people will laugh and accept them as a joke. Yet, are they? They are not correct when we apply 'common sense' and listen to our 'gut feelings.' So from a logical point of view they are not logically acceptable statements.

Common Sense and logic

Finally, I would like to present another mathematically logical problem. This was a question asked by President A. Lincoln:

> If I should call a sheep's tail a leg,
> how many legs would it have?
>
> "Five." [says someone]
> 'No, only four, for my calling the tail a leg would not make it so.'

This is an example of logic being used correctly. Just because someone calls a tail a leg, it isn't. It is still a tail!

When working with logic problems, common sense should be the rule-of-the-day. As President Lincoln said, "calling the tail, a leg, would not make it so!'

> *Note*:
> The actual quotation comes from *The Weekly Standard* [Raleigh, North Carolina], 29 October 1862.) Here is the quotation as found in the paper's article:
>
> > "OLD ABE GETS OFF ANOTHER JOKE. -- A couple of Abolitionists having called upon Old Abe to persuade him to issue his Emancipation Proclamation -- that is, before he issued it -- he got off the following good thing and knock down argument against his own act: "You remember the slave who asked his master -- if I should call a sheep's tail a leg, how many legs would it have? 'Five.' 'No, only four, for my calling the tail a leg would not make it so.' Now, gentlemen, if I say to the slaves, 'you are free,' they will be no more free than at present."

What is logic?

The word logic comes from the Greek word *logos*, translated to mean: "sentence", "reason" and "rule." Using this knowledge, we could research the official meaning of what logic is and see definitions like:

Wordnetweb.princeton.edu defines logic as –
- the branch of philosophy that analyzes inference
- reasoned and reasonable judgment; "it made a certain kind of logic"

merriam-webster.com/dictionary defines it as –
  (1) a science that deals with the principles and criteria of validity of inference and demonstration : the science of the formal principles of reasoning

Both official definitions of the word *logic* must be correct. However, many people look at the above definitions and have no idea what they mean! Rather than trying to define logic by an official definition, think of logic as *the study of the principles of correct reasoning*. Logic is concerned with the principles of *correct* reasoning not the *psychology* of reasoning! That is important and brings us back to using 'common sense' when reasoning out a problem.

A few 'easy' logic problems

If you think through a word problem, keeping common sense in the back of your mind, most can be solved without the need for writing them down. Exercise 10 has three logic problems that you should be able to easily figure out. They may take you a while. Just take your time and use your common sense.

EXERCISE 10

The first two problems are based on the premise that you are running in a race …

1. In the first race, you are passing the competitor who is in 2nd place. What's your position after passing?

2. Then, you race again and you overtake the competitor who is in last place. What's your position afterwards?

- Answers are on the next page -

For the third problem, you need to do a little addition. Please try to do this in your head – without a calculator:

3. Take 1000 and add 40 to it. Add 1000 to that. Add 30. Now add another 1000 and then add 20. Add 1000. Finally add 10. What is the total?

- Answer is on the next page -

Strategies for solving logic problems

Logic problems can be very rewarding. They can offer a mental challenge. You may find yourself working on one and trying to figure out why it seems so difficult. But, once completed, there is a feeling of accomplishment; hungering for more.

When tackling a logic problem it is a good idea to have a plan of attack. Here are a few ideas, to build a strategy for solving them:

1. Read the whole problem. During the first reading don't try to look at the specifics; rather, try to understand the overall idea about the nature of the problem.
2. Now, re-read the problem. This is the time to look at the particular information given in the problem. Pay attention to "What you are being asked to do?" – "What should the answer look like?"

ANSWERS: (Exercise 10) Easy Logic
(1) You are now in second (2nd) place. The person you passed was behind the first place runner, so you moved from third to second place.
(2) There are two answers to this problem:
    a. Since the person you are passing is in last place (the last person in the competition) it is possible to pass that person by being in any other position and running so fast that you have come around to the last person (running way out past them). So if you were in 3rd place, you would still be in 3rd place and be the 3rd person to pass the last place person …. The two people in front of you would have passed the last place person before you.
    b. The second possible answer is that, you can't overtake the last person who is in last place. Of course it is possible if you overrun the track and come upon him as in answer (a).
(3) 4100. Many people answer this incorrectly saying 5000. This is common due to the nature of human thought and rising a number (carrying over) to the next highest number. Counting by 1000 and then tens brings the person to a point that they need to carry from tens (90) to 1 hundred; however, they forget to raise to 100 and jump to raising by 1000 from 4090 to 5000. But the correct answer is 4100.

Once you have read and re-read the problem you should be ready to work on the problem. To work on the problem, consider these continuing steps:

3. Create a list, series of lists, or a model. Most word problems offer more than one item to consider in solving the problem. You may keep this list in your head or put it on paper.   While going through the clues you may need to create a table and put the information in it. Any specific information found should be put in the table.
4. Work backwards; if necessary. Once information has been placed in a list or table, it is often easier to work backwards to eliminate issues.
5. Decide if your answer makes sense. The last thing to do is to compare your answer with the original problem. Make sure that your answer satisfies each of the clues in your problem.

The above five steps will help you solve any logic problem. Just remember the key to all logic problems is to use common sense.

Some more logic problems
Here are a few more logic problems that will make you think.

EXERCISE 11
1. If two days before tomorrow is Friday, what day is it today?
2. If there are 22 horses and 8 chickens in a barn, how many legs are there between all the farm animals?
3. Mary's mother has four children. The first one is named April; the second is named May, and the third one June. What is the name of the fourth child?
4. You have a five gallon measuring device and a three gallon measuring device. Using just these two devices, you need to measure out exactly four gallons.

- Answer is on the next page -

ANSWERS: (Exercise 11) More logic problems
(1) <u>Saturday</u> – two days before tomorrow is. Starting with Friday move FORWARD two days – Fri, Sat, Sunday. Sunday . So Sunday is tomorrow; thus today is Saturday .
(2) <u>104</u>. There are 22 x 4 horse legs and 8 x 2 chicken legs – thus 88 +16 = 104.
(3) <u>Mary</u>. It starts with Mary's mother has four children … so Mary is one of them.
(4) This takes several steps. First fill the five gallon container. Fill the three gallon container from the five gallon container; leaving two gallons in the five gallon container. Empty the three gallon container. Put the remaining two gallons in the five gallon container into the three gallon container. Fill the five gallon container one more time. Pour one gallon from the five gallon container into the three gallon container, filling the three gallon container. Now you have *four gallons in the five gallon container.*

Two difficult problems in logic

Up to this point, most of the problems have been fairly easy to solve; although some may have taken more time to answer than others; here are two tough ones:

EXERCISE 12

<u>*Problem 1*</u>: There are three coins like the three shown:

One coin has heads on both sides; one tails on both and one is a normal coin with a head and tail.

All three coins are in a bag and you draw one of the coins out – without looking at it. Now you flip it and uncover it. The coin shows HEADS up – *What are the odds the other side will be a head?*

<u>*Problem 2*</u>: There are two towns at the end of a forked road. One is the town where everyone tells the truth; the other is where everyone always tells a lie. The truth people live in TRUTH town, and the lying people live in LIE town.

You follow a path on your way to see a friend in TRUTH town. You come to a fork in the road and there is no sign! One way goes to TRUTH town and the other to LIE town. There is a person sitting on a rock at the fork. He is from one of the town's. What one question can you ask him and then know which way leads to TRUTH town?

ANSWERS: (Exercise 12)
Problem 1: There are three head conditions; therefore the odds of a head on the other side is 2 out of 3
Problem 2: Show me which path I follow to your town!

End Game for the Day

Today's End Game is to create a shirt by folding a United States Dollar bill. Folding paper is known as Origami. Origami is the Japanese word for paper folding. *"Oru"* means to fold and *"kami"* means paper.

> ### *Note*:
> Origami actually began in Japan with the introduction of paper from China. The art of paper making originated in China in the year 102 (A.D./C.E.) The secret for making paper was a closely held art in China for over two hundred years; finally making its way to Japan and Korea. It is said that a Buddhist monk carried the secret art to Japan. Paper quickly became a part of the lives of the people of Japan. They introduced silk threads and colors to the paper and began using it for folding into different shapes. Gifts were decorated with *"noshi."* Noshi is actually a folder wrapper, often looking like a flower, that is attached to a gift as a way of saying "good wishes."

Creating a money shirt

Would you like to use a dollar and make a money shirt like the one here?

It's easy! Just follow the ten steps below (It's best to start with a clean, crisp bill and folds should be sharply creased (go over the fold with a fingernail on a flat, hard surface.)

Step 1:

The first thing you will need is a dollar bill to use. So find one to work with.

Step 2:

Start by folding the bill exactly in half lengthwise (long ways). Try to fold toward the front of the bill. It will make a nicer design.

> ### *Note*:
> For your first attempt, it will be easier to follow the exact same bill orientation as shown in the photos.

Step 3:

Un-fold the bill you folded in half during step 1. Position it face side up and Washington's head upside down. The fold created a crease along the center of the dollar and will be used in later steps.

Step 4:

Washington nose is the point!
With Washington's face visible, fold from the end at the back of his head inward so that the edge of the dollar goes up to Washington's nose.

Your dollar should now look like the one here.

Step 5:

Using the crease you made in in Step 1, fold each half of the bill inward, so the edges meet in the middle at the crease from fold 1;

Step 6:

Start at the end that you folded toward Washington's nose. Make 2 angle folds so the corners extend out approximately 30 degrees. They should look like an open collar on a shirt. The bottom of each new fold is near the "O" of the word Dollar.

Step 7:

This is the second easiest step. All you need to do is flip the bill over, so that it looks like the one here. Those shirt collars should now appear to be shirt sleeves.

Step 8:

Go to the opposite end of the bill – away from the sleeves. Fold the white (unprinted) edge of the bill over onto itself (near the seal). The unprinted part should meet the printed part. A split will be on this side of the dollar; once folded. (See picture.)

Step 9:

Flip the bill over again. Go to the edge of the bill you just folded (the white can't be seen) and fold each side from the outer edge towards the middle. To help, pinch the white edge in at the center of the top; to create roughly a 30 degree angle. (the collar should look like the picture)

The other end will become the sleeves.

Step 10:

All that is left to do is to fold the bottom of dollar (sleeves) upward, so the bottom edge (sleeves) goes under the "collar."

Your shirt should look like the one here as you fold in the middle.

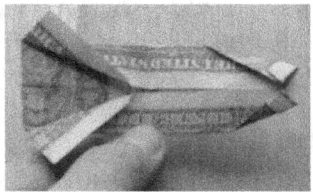

Just fold the bottom part up and place it under the collar.

Once finished the shirt should look like this:

Now this is a good looking shirt! = = = >>

You can create shirts from other USD bills. If you want to have some color in your shirts, try a $5.00 or $10.00 USD bill. You can even create a shirt with a necktie.

To learn more go to *Yahoo* or *Google* and enter the keywords: "dollar bill shirt and tie instructions"

Day Three

WELCOME TO DAY 3 OF MATH 4 2-DAY! The journey continues. If you need to review any material covered during days one or two, take the time to go back and review any part in this book.

Today we will focus on the Laws of Math and the topic of fractions.

Today's Outline

Quotations for the Day
Bit of Fun in Math
  Murphy's Law in Math
Laws of Math
  Mathematical Properties
  Order of Precedence
Mathematical Curiosities (Part 2)
  Prime Numbers in the world of math
  Sieve of Eratosthenes
Fractions – the Math of Parts
  Least Common Denominator and Greatest Common Factor
  Fraction Math
Calculator Tricks (Part 3)
Math is Logic – or – The Logical Methods of Math (Part 2)
Simple Math is All You Need (Part 3)
  Fun with multiplying by 12, squaring by numbers ending in 5 and 0
  Multiplying two numbers near 100, x99, x101, x1001, and two teen numbers
Endgame for the Day
  Dots and Thinking outside of the box

Objectives:

(1) Build on the Laws of Math
(2) Continue to work with ideas in the field of Logic
(3) Build confidence when finding prime factors
(4) Continue to perform mental math – avoiding the calculator
(5) Expanding our mental skills in math while thinking abstractly

Quotations of the Day

ℵ · ℜ

A Mind is fire to be kindled,
Not a vessel to be filled!
-- *Plutarch*

Mestrius Plutarchus (c. 46 - 120 A.D.) also known in English as *Plutarch*, was a Greek historian, biographer, and essayist. He was once a Roman Empire priest at the temple of *Apollo of the Delphic Oracle*. He wrote in a language known as *Attic Greek* (known by the Scholars of the day.) His best known work is titled *Parallel Lives* – it is a collection of biographies of several Greeks and Romans.

ℵ · ℜ

Mathematics is the science of what
is clear by itself!
-- *Carl Jacobi*

Carl Jacobi (c 1804 - 1851) was a Prussian (modern day Germany) mathematician, who is known as one of the most famous teachers of his time and considered to be one of the greatest mathematicians of all time. In 1829 he wrote his classic treatise on Elliptic Functions; they are a critical concept in mathematical physics and number theory.

ℵ · ℜ

Errors using inadequate data are much
less than those using no data at all!
-- *Charles Babbage*

Charles Babbage (c 1791 - 1871) is known as the "Father of Computing" for his contributions to the design of the computer through his *Analytical machine* and *Difference Engine* (special purpose device to produce tables.) He also invented the cowcatcher on trains, uniform postal rates, osculating lights for lighthouses, and the ophthalmoscope.

ℵ · ℜ

Bit of Fun in Math

The topic for today is Murphy's laws and its impact on the field of mathematics.

There are many laws for mathematics. Some deal with operations (addition, subtraction, etc.), properties, order of precedence and even others with higher order issues, like logarithms and calculus. But there is also a law known as Murphy's Law. It simply states,

"If anything can go wrong, it will."

And, we can probably assume, that it will do so,
at the worst possible time, causing the most damage!

Here are a Baker's dozen (13) ways of how Murphy's Law applies to Math.

Murphy's Law in Mathematics

1. The harder you study, the farther behind you will get and the less you know!
2. Every problem is harder than it looks and takes much longer than you expected it to take!
3. To solve any problem; it really helps to know the answer!
4. Knowing mathematics and teaching mathematics are not necessarily equivalent! [Knowing Math ≠ Teaching Math]
5. Mathematical Proofs won't convince anybody of anything. People need to see them work before they start to believe!
6. What is "obvious" to everyone else won't be "obvious" to you!
7. The notes you took in class and understood; transformed themselves into hieroglyphics by the time you got home!
8. Textbooks are obviously written for people who already know the subject!
9. Every math idea in the math book is expressed in incomprehensible, mind-blowing terms!
10. The answers you need are never in the back of the book! (The even numbers)
11. No matter how much time you spend studying for exams, it will never be enough!
12. The problems you already know the answers to won't be on the test; the ones you struggle with will always be on it!
13. The answers to problems you couldn't solve during the exam will become obvious *after* you hand in your paper!

Laws of Math

We had some fun earlier with Murphy's Law as it applies to Mathematics. This section deals with some basic methods, or rules, that apply in mathematics. The two primary areas covered are math properties and order of precedence.

Mathematical Properties

When working in math there are a few properties related to operations. Of course the first issue is the operations themselves.

Operations

An *operation* is the workhorse of numbers. Operations change numbers. The word operation comes from the Latin *operari*, meaning "to work." There are six primary operations that actually "work on" numbers:

They include:

Addition (+)	Subtraction (−)
Multiplication (**x** or ● or *)	Division (÷ or **/**)

Exponentiation – raise to a power of (^x or ^{x})
Grouping – use of parenthetical expressions like – ( ) or [] or other symbols

These six operations are used every day to perform mathematical calculations.

> *Note*:
>
> Using these operations for solving mathematical problems is known as *Algebraic Notation*.
>
> There are also two special operators: *Square root* – ($\sqrt{}$) and *Factorial* – (!) which will be covered another day. Both are used to perform special operations.

Associative Property

The associative property *associates* two or more terms (numbers) together creating one unit. Parentheses are used to *group* or *associate* two or more terms (in an operation of three or more terms.) The grouping, parentheses, is an essential part of the associative property; the actual grouping '*associates*' terms together.

The parenthesis can be placed anywhere in a group of terms, in any way, and will not change the outcome of the final answer.

This property only applies to addition and multiplication operations of three or more mathematical terms, using the same operator.

> *Remember*:
>
> No matter how you combine them, the answer will be the same!

Addition and Multiplication are associative. The answer will always be the same, no matter how you combine like term objects (a, b, and c), when using addition or multiplication. Using variables the associative property can be defined as:

$$a + (b + c) = (a + b) + c \qquad \text{(Addition)}$$

$$a * (b * c) = (a * b) * c \qquad \text{(Multiplication)}$$

You can put some numbers into the above examples to see how the associative property works:

$3 + (4 + 6) = 13$ AND $(3 + 4) + 6 = 13$ $8 + (1 + 5) = 14$ AND $(8 + 1) + 5 = 14$

$3 * (2 * 6) = 36$ AND $(3 * 2) * 6 = 36$ $8 * (2 * 5) = 80$ AND $(8 * 2) * 5 = 80$

No matter where the parentheses (groupings) are placed in the formula, the answer will not change.

It is important to show that you can NOT mix operators using division and subtraction. The associative property does not apply to these two operators. NEVER mix Add/Subtract or Multiply/Divide together – they just aren't associative:

$$a - (b + c) \neq (a - b) + c$$

A practical example is $10 - (5 + 3) = 2$ where $(10 - 5) + 3 = 8$

$$(a / b) * c \neq a / (b * c)$$

A practical example is $(10/5)*3 = 6$; $10/(5+3) = 1.25 \ldots$ and $\ldots$ $6 \neq 1.25$

The associative property applies to addition / multiplication. It is the grouping, by parenthesis, groups of terms, without affecting the solution to the problem.

Commutative Property

The commutative property is one where a mathematical operation can change the order of the terms (numbers) without affecting the results.

Think of the root word of "*commutative*" as "commute." Commute means to "move around" or "change the order or arrangement"; so the Commutative Property is the one that refers to moving stuff around.

Formally, it is the ability to *change the order* of the numbers involved in a series of like operations without changing the result.

The commutative property applies to addition and multiplication only; like the associative property. No matter how you combine like term objects (a, b, and c) using addition or multiplication, the answer will always be the same.

Using variables *a*, *b*, and *c*, the commutative property can be demonstrated through addition and multiplication.

For addition, the rule is:

$$a + b = b + a \qquad \text{or} \qquad a + b + c = b + c + a = a + c + b$$

You can replace the variables in the above examples with actual numbers to demonstrate the commutative property for addition:

$$5 + 2 = 2 + 5 \;(7) \qquad \text{or} \qquad 2 + 3 + 4 = 3 + 4 + 2 = 2 + 4 + 3 \;(9)$$

For multiplication the rule is:

$$a * b = b * a \qquad \text{or} \qquad a * b * c = c * b * a = a * c * b$$

Replacing the variables with values you can see how the commutative property works for multiplication:

$$4 * 5 = 5 * 4 \;(20) \qquad \text{or} \qquad 2 * 3 * 4 = 4 * 3 * 2 = 2 * 4 * 3 \;(24)$$

Remember, no matter how you move the variables around or change their order, the resulting solution remains the same. This is what the commutative property does; it allows you to commute, or change the order of the individual variables, without affecting the answer – it remains the same.

Again, like the Associative Property, the Commutative Property does *NOT* apply to Subtraction and Division. Here are a couple of examples that demonstrate how it doesn't apply to either operation:

$$2 - 1 \neq 1 - 2 \text{ (subtraction)} \qquad 3 / 4 \neq 4 / 3 \text{ (division)}$$

An interesting observation can be made for mixing addition with subtraction *OR* mixing multiplication with division; when discussing the commutative property. Here are a couple of examples to demonstrate mixing addition with subtraction and then mixing division and multiplication: (no associative property is assigned)

ADDITION with SUBTRACTION
$$a - b + c = c - b + a = a + c - b = -b + a + c$$
$$5 - 2 + 3 = 3 - 2 + 5 = 5 + 3 - 2$$

MULTIPLICATION with DIVISION
$$a * b / c = a / c * b = b / c * a$$
$$2 * 4 / 8 = 2 / 8 * 4 = 4 / 8 * 2$$

So, although subtraction and division are *NOT* commutative, you can mix addition with subtraction or multiplication with division and change the order; not affecting the answer. However, these mixed operations do not fall under the commutative property since it only applies to addition and multiplication. They are just an interesting observation.

Distributive Property

The distributive property is one where "multiplication *distributes* over addition." It is used to remove the parenthesis from a formula by separating or breaking it into its individual parts, without affecting the result. In other words, it makes working with numbers easier by expanding the problem into separate parts.

To distribute means to give or provide pieces of something to many. In math, it refers to applying one operation over another. You distribute multiplication over addition or subtraction.

The distributive rule of multiplication over addition is:

$$a (b + c) = (a*b) + (a*c) = ab + ac \qquad a (b - c) = (a * b) - (a * c) = ab - ac$$

You can replace the variables in these examples to demonstrate:

$$
\begin{array}{rcl}
5 * (2 + 3) & = & (5 * 2) + (5 * 3) \\
5 * 5 & = & 5 * 2 + 5 * 3 \\
25 & = & 10 + 15
\end{array}
\qquad
\begin{array}{rcl}
5 * (5 - 3) & = & (5 * 5) - (5 * 3) \\
5 * 2 & = & 5 * 5 - 5 * 3 \\
10 & = & 25 - 15
\end{array}
$$

So the distributive property lets you remove the parenthesis from a formula by breaking the two parts of the formula (multiplication over addition) into their expanded parts; simply multiplying each part of the addition problem by the multiplicand and adding them together.

Note:
Subtraction is actually adding a negative number to another number, e.g. $10 - 4$ is $10 + (-4)$

Order of Precedence

Also known as the Order of Operations

Before covering this topic, a simple math problem is needed to demonstrate how we solve problems. What is the answer to this problem?

$$2 + 3 * 4 = \underline{\hspace{2cm}} ?$$

How do you solve the problem with two different operations – addition and multiplication? Is the correct answer, *20* or *14*? In theory, both are possible. The correct answer depends upon how you look at the problem. For example-

Solving from left to right, we get: $2 + 3 = 5$ then $5 * 4 = 20$
Solving by using an order of operations, we get: $3 * 4 = 12$ then $12 + 2 = 14$

Solving math problems are confusing if we can't be sure of the answer; or worse, if the same problem can be solved different ways, resulting in different answers.

So which is correct? The correct answer, as shown, is the second choice – 14.

Why do we need an order of operations?

Using the previous problem, it is obvious why we need to have a universal system for solving math problems. Of course, the problem would have been much clearer if parenthesis had been used to specify the order needed to solve it.

For emphasis, if the addition part of the problem had been grouped (enclosed between parentheses) you would solve the addition operation first; then the correct answer would be 20. Of course, had the multiplication operation been in parenthesis, the answer reverts back to 14! So parentheses can be used to clear up any confusion when solving math problems.

What do you do if there are no parentheses? Well, that is the problem, isn't it? Is there some 'magic' order of operations that we must do, so that we all arrive at the same answer? Actually, the answer to that question is YES.

At some point, everyone agreed, for the sake of clarification, some defined method was needed. It would insure that everyone performing math calculations with mixed operators would obtain the same consistent *correct* answer.

Although we don't know the actual origin of the who or when a specific order of operations was created, it was adopted at some point and has been used ever since.

According to the website, *Math Forum's Ask Dr. Math*, the order of operations probably began after algebraic notation (an operation sign placed between numbers) was created. Concerning the history of operations order, Dr. Math says –

> "... the basic rule (that multiplication has precedence over addition) appears to have arisen naturally and without much disagreement as algebraic notation was being developed in the 1600s and the need for such conventions arose. Even though there were numerous competing systems of symbols, forcing each author to state his conventions at the start of a book, they seem not to have had to say much in this area. This is probably because the distributive property implies a natural hierarchy in which multiplication is more powerful than addition, ..."

By creating a specific method, or order to implement, when doing math calculations, everyone *should* arrive at the same answer.

The Order of Precedence/Operations

An order of precedence or the order of operations simply refers to the order that should be used when performing math operations. In America we use an *acronym, PEMDAS* to represent that order. The order of precedence for mathematical operations is performed, from *left to right* and in the following order:

Operations within **P**arenthesis
Exponentiation (raising to the power of – e.g. x^2)
Multiplication and **D**ivision (whichever comes first)
Addition and **S**ubtraction (whichever comes first)

Note:

An *acronym* is a word formed from the first letters of several words. In the case of the word PEMDAS, the <u>P</u> is from the word **P**arenthesis, <u>E</u> is from **E**xponentiation, <u>M</u> is from **M**ultiplication, and so on.

Using this order, the person solving a multiple operator problem would first look for operations in parenthesis. Then proceed to any exponentiation. Following these by performing any multiplication *or* division they find (from left to right); finally, solving addition *or* subtraction (again from left to right). Here is a complex problem and the steps used to solve it:

$$(3 + 6) / 3 * 4 + (2^3 - 2) + 23 * 5$$

Steps to solve –

1.	$(3 + 6) = 9$	$9 / 3 * 4 + (2^3 - 2) + 23 * 5$
2.	$2^3 = 8$	$9 / 3 * 4 + (8 - 2) + 23 * 5$
3.	$(8 - 2) = 6$	$9 / 3 * 4 + 6 + 23 * 5$
4.	$9 / 3 = 3$	$3 * 4 + 6 + 23 * 5$
5.	$3 * 4 = 12$	$12 + 6 + 23 * 5$
6.	$23 * 5 = 115$	$12 + 6 + 115$
7.	$12 + 6 + 115 = 133$	ANSWER is 133

First find any operations inside parentheses. The first set of parenthesis is simple addition $(3 + 6)$. Then the second operation in parenthesis is one with a 2 to the power of 3 inside it. Before solving the problem the 2 to the power of 3 must be solved, then subtract 2 from that answer. Then (from left to right in the equation) the division and multiplication parts were solved: 9 divided by 3, then 3 times 4, and then the last multiplication of 23 times 5. Finally all the partial answers (from the above operations) were added together.

That's all there is to PEMDAS. Remember: multiplication and division are equal; as are addition and subtraction. When solving for either, start with the left most multiply/divide or add/subtract and move to the right solving each one as you come to them.

Remembering the acronym PEMDAS

Some people remember order of precedence through use of a phrase constructed by words made up from the first letter of each operation, e.g. PEMDAS. Follows are several examples. The first is the 'traditional' one found in many math books.

- Please Excuse My Dear Aunt Sally
- Pink Elephants Make Dandy Apple Sauce
- People Everywhere May Drink Any Soda
- Pandas Eat More Dumplings And Stickers
- Pills End My Disease And Sickness

An interesting tidbit, PEMDAS is not the only *'Order-of-Precedence'* acronym. Other countries use other acronyms. For instance, Canada's acronym is BEDMAS; a mnemonic for Brackets, Exponents, Division, Multiplication, Addition, and Subtraction. In the UK, Australia, New Zealand, and South Africa, the acronym BODMAS is commonly used for Brackets, Orders, Division, Multiplication, Addition, Subtraction. Since multiplication and division are the same rank, it is sometimes written as BOMDAS, BIDMAS or BIMDAS where the "I" stands for Indices.

> *Note*:
>
> In Danica McKellar's book *Kiss My Math*, there is a new catch phrase: "*Pandas Eat: Mustard on Dumplings, and Apples with Spice.*" This one is suggested for use as that the intention is that *Mustard and Dumplings* is a "dinner course;" while *Apples and Spice* are a "dessert course." By using this analogy, the order of precedence becomes a non linear string of operations. Instead of thinking that multiplication MUST be performed before division or addition before subtraction, thinking of the "dinner course" operations can be done in any order and are considered together and performed left to right. Thus the addition and subtraction are also considered together, again performed in order from left to right.

Does the order make a difference?

Do you remember the commutative law of multiplication – $a \times b = b \times a$; and the associative law of multiplication: $a \times (b \times c) = (a \times b) \times c$? Are these laws always true?

this question comes up a lot in higher order mathematics. It is especially true when talking about *quaternion* numbers which violate the commutative laws of math or *Cayley* numbers (*octonions*) that violate the associative laws. This class is not intended to cover these topics, you should be aware that there are a few exceptions. Even in calculus, the order of addition is critical when working with *convergence*. Specifically, there are number sequences that fall into absolute and conditional convergent states; in conditionally convergent states, the order is critical.

Again, it is also true when using calculators or computer programs to perform complex calculations. Understanding how these programs execute your expressions is essential. For instance, *Mathematica* uses a unique way to calculate expressions.

Here are two mathematically identical problems:

$$1 - 1 + 10\wedge\text{-}20 \qquad \text{and} \qquad 1 + 10\wedge\text{-}20 - 1$$

In *Mathematica* they produce different answers. In the first one, subtraction is performed first and the result is zero; adding 0 to $10\wedge$-20, results in $10\wedge$-20.

In the second expression, addition is performed first. With only 16 digits of mantissa (using *Mathematica*), the number $10\wedge$-20 is lost (results in a calculation of 0), during the round off, and the resultant answer is 1. Then the subtraction of 1 gives an answer of zero. (This demonstrates that floating-point arithmetic need not satisfy the associative rule.)

Mathematical Curiosities (Part 2)

This section will discuss prime numbers, define what prime numbers are, and offer a method for finding all the prime numbers within a given (*composite*) number.

There is an ancient way, known as the *Sieve of Eratosthenes*, which can be used to determine prime numbers – especially from 1 to 100.

Once you learn how to find prime numbers a method known as *Prime Birthday Cake Factoring* will be introduced.

Prime Numbers in the world of math

A prime number is a natural or counting number (1, 2, 3, and so on) *greater than* 1 that can be divided *evenly* only by itself and 1. The first few prime numbers are 2, 3, 5, 7, 11, 13, 17, 19, 23, 29, and on and on. Again, *one* is not a prime number.

There is a song "One", from Three Dog Night's first album *Three Dog Night* (1969). It talks about the fate of ONE:

"♫ ♪ ♫ One is the loneliest number that you'll ever do. ♫ ♫♪"

These words say it all about the number 1. The number 1 couldn't possibly be a prime number because it is all alone. It could be divided by itself but to also divide it by 1 is redundant – thus it doesn't fit the rule; so "one is the loneliest number"! Of course, the real reason why 1 is not a prime number is merely convenience.

If 1 were prime it would cause a problem with determining factoring of numbers for their primes. For example, if 1 was prime then the prime factorization of 6 would not be unique since 2 times 3 actually would equal 1 times 2 times 3. A number that can be written as a product of prime numbers is *composite*. Thus there are three types of natural numbers: primes, composites, and 1.

Composite Numbers:
A composite number is one that can be divided evenly by other numbers. It is the product of two smaller numbers (multiplying two smaller numbers together to get the number.) Most numbers are composite numbers – 4 is a composite number since it can be divided by 2, 1 and itself. 27 is a composite number that can be divided by 9 and 3, as well as 1, and itself.

Prime numbers and their properties were first studied extensively by the ancient Greek mathematicians.

The mathematicians of Pythagoras' school (500 BC to 300 BC) were interested in numbers for their mystical and numerological properties. They understood the idea of primality (prime numbers) and were interested in *perfect* numbers.

Around 200 BC the Greek, Eratosthenes, devised an *algorithm* for calculating primes; known as the *Sieve of Eratosthenes*.

Sieve of Eratosthenes

Eratosthenes, a Greek scholar and mathematician of the third century BC (275-194 BC/BCE) devised a 'sieve' to discover prime numbers.

> *Note*:
> A sieve is similar to a strainer that you drain spaghetti through when it is done cooking. The water drains out, leaving your spaghetti behind.

Eratosthenes' sieve drains out composite numbers and leaves prime numbers behind. To use the sieve of Eratosthenes for finding prime numbers up to 100, make a chart of the first one hundred whole numbers (1-100), like this one –

1	2	3	4	5	6	7	8	9	10
11	12	13	14	15	16	17	18	19	20
21	22	23	24	25	26	27	28	29	30
31	32	33	34	35	36	37	38	39	40
41	42	43	44	45	46	47	48	49	50
51	52	53	54	55	56	57	58	59	60
61	62	63	64	65	66	67	68	69	70
71	72	73	74	75	76	77	78	79	80
81	82	83	84	85	86	87	88	89	90
91	92	93	94	95	96	97	98	99	100

Now follow these steps -

1. Cross out 1, because it is not a prime number.
2. Circle 2, because it is the smallest positive even prime. Now cross out every multiple of 2; in other words, cross out every second number.
3. Circle 3, the next prime. Then cross out all of the multiples of 3; in other words, every third number. Some, like 6, may have already been crossed out because they are multiples of 2.
4. Circle the next open number, 5. Now cross out all of the multiples of 5, or every 5th number.

Continue doing this, circling the 7 next, until all the numbers through 100 have either been circled or crossed out. The circled numbers are all the prime numbers from 1 to 100! You should have the following numbers:

2	3		5		7		11		13		17		19		23		29		31		37		41
43	47		53		59		61		67		71		73		79		83		89		97				

> *Note*:
> The only prime number found to date that is even is the number 2!?

Fractions – the math of parts

Today we will work with fractions and explain their importance in the world of mathematics. This section may be nothing more than a review. If you are already familiar with fractions, you can skip it or simply review the material.

Fractions are an important concept first introduced in Middle school. They can be challenging for some students. Many reports show that fractions are one of the most difficult topics for children to master (2001 National Assessment of Educational Progress – NAEP report.) Part of the problem lies within the meaning of fractions. There are several meanings; fractions are:

- Parts of a whole: an object is divided into *"b"* parts, then a/*b* means *"a of b"* parts
- The size of a portion when an object of size *"a"* is divided into *"x"* equal portions
- The quotient of the integer *"a"* divided by *"b"*
- The ratio of *"a to b"*
- An operator: that carries out a process, such as *"3/7 of"*

Least Common Denominator and Greatest Common Factor

A concept needs to be discussed before working with fractions and addition/ subtraction math. If two fractions have different denominators (bottom numbers), you need to understand *Least Common Denominator* and *Greatest Common Factor*.

What is a Least Common denominator?

The *Least Common Denominator* (LCD) is a single number that each fraction's denominators (bottom numbers) can evenly divide into – think of it as the *Least Common Multiple* (LCM). In fact, the term LCM and LCD are often interchanged. For example, to do addition/subtraction with the numbers 1/9 and 1/12 you need to find the LCD/LCM that the numbers 9 and 12 both divide into.

> <u>Note</u>:
> The *Least Common Multiple* is a number that can be evenly divided by two or more numbers.

Finding Least Common Denominator

One way to find the LCD is to create a table of multiples for each number – 9, 18, 27, 36, 45, 54 and 12, 24, 36, 48, 60. Looking at the tables you can see the first number that is common to both numbers is 36 –the LCD/LCM!

Another way is to look at the two numbers. Both can be divided by 3. If you then multiple 9 x 12 you will get 108. Since both can be divided by 3 you can divide 108 by 3 and also arrive at 36. Both numbers 9 and 12 go into 36 evenly; 36 is the LCD.

What is the Greatest Common Factor?

The official definition for the *Greatest Common Factor* (GCF) is something like, "the highest number that divides exactly into two or more numbers." Determining the GCF of two numbers is a great way to simplify fractions like 12/30.

To better understand what a GCF actually is; just take a look at the three words:

Factor –factors are the numbers you multiply together to get another number:
> 2 (factor) times 3 (factor) equals 6

Common Factors – are those factors that are found in several numbers (2 or more).
> If you determine all the factors for 12 and 30 you will get a list like the following:
>> 12 – 1, 2, 3, 4, 6, 12
>> 30 – 1, 2, 3, 5, 6, 10, 15, 30
> Looking at these lists, the common factors of 12 and 30 are 1, 2, 3, and 6.

Greatest Common Factor – is the largest of the common factors between two or more numbers
> 6 is the Greatest common factor between 12 and 30 (of 1, 2, 3, and 6).

> <u>Note</u>:
> Some people believe that the GCF is only usable for finding common factors. However, knowing the GCF makes the process of finding the LCD/LCM easy.

Before discussing the Greatest Common Factor (GCF) you need to understand how to find all prime *number factors* for any number. By breaking a number into its prime factors (a x b x c = N) you can easily find the GCF between two numbers.

Prime Birthday cake factoring

First you need to find the prime factors of any number. Traditionally factoring is performed by creating a factor tree. You can also find factors of any given number using a method known as the *Prime Birthday Cake* method. It is a series of divisions that will form a birthday cake like structure.

The following demonstrates creating a factor birthday cake with a candle (number 1) for the number 12.

1. Divide 2 into 12 (at the bottom of the calculation) place 6 above	1
2. Divide 2 into 6 and place 3 above it	3 | 3
3. Divide 3 into 3 and place 1 above it	2 | 6
	2 | 12

By extending the lines from each division symbol like these, it should look like a birthday cake!	1
	3 | 3 |
The numbers on the left (divisors) are all the prime factors of the	2 | 6 |
number 12 (2,2,3). Remember 1 is not a prime number – (thus it creates the candle)	2 | 12 |

Here is simple exercise to find prime factors of a few numbers; using the *Birthday cake method* – 70, 60, 27, and 48. The first exercise already has one of the prime factors:

EXERCISE 13

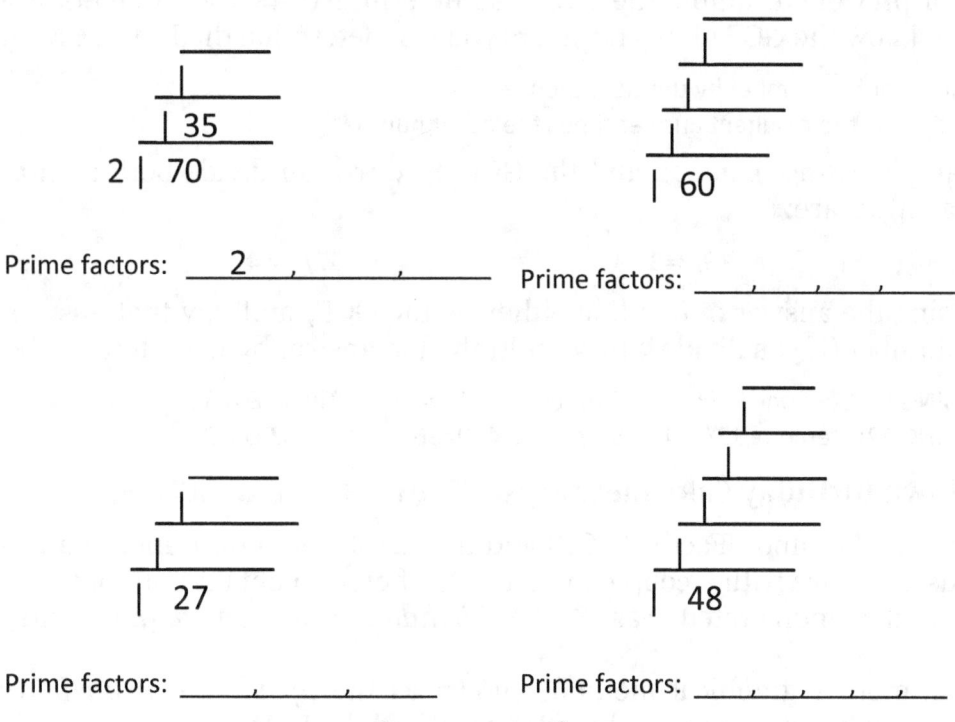

Prime factors: ___2___ , ___ , ___

Prime factors: ___ , ___ , ___ , ___

Prime factors: ___ , ___ , ___

Prime factors: ___ , ___ , ___ , ___

- Answers are on the next page -

Using the *Birthday cake* method to find all prime factors of a number is relatively easy and offers another graphical method for finding them. It uses division over and over until 1 (the candle) is reached. Remember 1 is not a prime factor.

<u>Note</u>:
Most students learn another graphical method known as building a prime factor tree. It is the universal method of teaching prime factorization taught today. Although a viable method, the *Birthday cake* method does seem more powerful.

This method for finding all prime factors of a number will be used later to find the GCF and to determine the Least Common Denominator of two numbers.

<u>Note</u>:
There are two additional ways to find the LCD and the GCF. They are explained later – the *upside-down birthday cake* method and the *Venn diagram* method.

ANSWERS: (Exercise 13) *Birthday cake method* for finding prime factors of a number
70 (2, 5, 7) – 60 (2, 2, 3, 5) – 27 (3, 3, 3) – 48 (2, 2, 2, 2, 3)

Using the GCF to find the LCD/LCM

One way to find the LCD/LCM is to work with the *Greatest Common Factor* (GCF). In a previous example the GCF was determined for the numbers 9 and 12 (36). If you know the GCF of two numbers you can determine the LCD/LCM by –

1. Dividing either number by the GCF; then
2. Multiplying the resultant answer times the other number

For example, using 9 and 12 and the GCF of 3, we can divide both by 3 and the resultant answers are:

$$9/3=3 \qquad \text{and} \qquad 12/3 = 4$$

Then using the answer of dividing either by the GCF; multiply that answer times the other number (if you divide 9 by 3 multiply that answer by 12). Here are both:

USING 9: Step one 9/3=3   Step two: 3*12 = 36   The LCD = 36
USING 12: Step one 12/3=4   Step two: 4*9=36   Again LCD = 36

Upside-down Birthday Cake method for finding GCF and LCD

The Greatest Common Factor (GCF) and the Least Common Denominator (LCD) were discussed above with a couple of methods offered to obtain both of them. Also discussed and demonstrated was *Prime Birthday cake factoring* to find prime factors.

There is another graphical method, known as the *upside-down birthday cake method* that can be used to obtain both the GCF and the LCD.

If you recall the Greatest Common Factor is the largest number that will go evenly into two numbers. For example, using the numbers 32 and 48, you can determine the greatest factor by writing down all combinations of factors for both numbers:

NUMBER	Combinations		Factors
32	1 x 32, 2 x 16, 4 x 8	=	1, 2, 4, 8, 16, 32
48	1 x 48, 2 x 24, 3 x 16, 4 x 12, 8 x 6	=	1, 2, 4, 6, 8, 12, 16, 24, 49

Writing down all factors for each number, you can quickly scan both lists of factors and see which numbers they have in common: 1, 2, 4, 8, and 16. The largest number each has in common is *16* – which is the GCF of 32 and 48.

Some people, using this method forget 3 x 16 = 48 and declare 8 as the GCF.

Although this is one way to solve for the Greatest Common Factor, it is prone to errors. If you forget one of the factors it could be disastrous!

The *upside-down birthday cake method*, for finding the GCF of two or more numbers, is more visual and many educators who teach it believe it is an effective and accurate method for determining GCF.

Using the Upside-down Birthday cake for finding the GCF

The best way to learn this method is to demonstrate it through a practical example. For this example you want to find the GCF for the numbers 32 and 48. To use this method, follow these steps:

1. Write the two numbers side by side on a single line with some space between them for example: 32   48

2. Draw a line underneath the numbers and another alongside (in front of) the leftmost number. This should form an upside down divide by sign. | 32   48

3. Think of the lowest prime number that can be divided into both? How about 2; place that in front of the process you built in step 2:   2 | 32   48

4. Now divide 2 into both numbers and place the answer below each number. Here is what your process should look like now:

   $$2 \mid \underline{\quad 32 \qquad 48 \qquad}$$
   $$16 \qquad 24$$

5. Repeat steps 2 through 4 until you have exhausted all possible prime numbers that will evenly divide into both numbers. Your problem should look similar to this:

   $$2 \mid \underline{\quad 32 \qquad 48 \qquad}$$
   $$2 \mid \underline{\quad 16 \qquad 24 \qquad}$$
   $$2 \mid \underline{\quad 8 \qquad 12 \qquad}$$
   $$2 \mid \underline{\quad 4 \qquad 6 \qquad}$$
   $$2 \qquad 3 \quad <\,< = = \text{STOP}$$

6. Now multiply all the numbers to the left of each step together: (2 x 2 x 2 x 2) to obtain the Greatest Common Factor of 16.

Looking closely at the calculations in step 5, it should look similar to an upside-down birthday cake – thus its name. Notice that the last two numbers, 2 and 3, have no number that will evenly divide into both; except 1.

Note:
Another name for this method is the Indian method. The author is unable to find the origin of the name or who first introduced this method to the world of mathematics.

Using the Upside-down Birthday cake for finding the LCD

Once you have used the *upside-down birthday cake method* for finding the GCF you can quickly determine the Least Common Denominator for two numbers. Start with the previous example, as shown below:

```
 2 \  32        48
    2 \  16        24
       2 \  8        12
          2 ) |  4        6
                 2        3   < < = = STOP
```

Notice that the example has an ellipse around the four prime numbers (2, 2, 2, 2) that are multiplied together to obtain the GCF for the numbers 32 and 48. Now that you have determined the GCF is 16 you are ready to use the same upside-down birthday cake to find the Least Common Denominator (LCD) for the numbers 32 and 48. To determine the LCD you look at the bottom/last line of the example:

```
     2        3    < = = Bottom/last line
```

Multiply these two numbers, 2 and 3, (circled above) together to get 6. Now simply multiply the GCF (16) times this resultant answer (6) and the LCD is 96.

To rephrase the process, follow these steps:

1. Solve for the GCF, (16 in the above example)
2. Multiply the two remaining numbers on the bottom line (2 x 3 = 6 in the above example)
3. Multiply the results of step 2 by the GCF (6 x 16) to get the LCD (96)!

It is time to do a couple exercises, using the Upside-down birthday cake method, to find the GCF and LCD:

EXERCISE 14

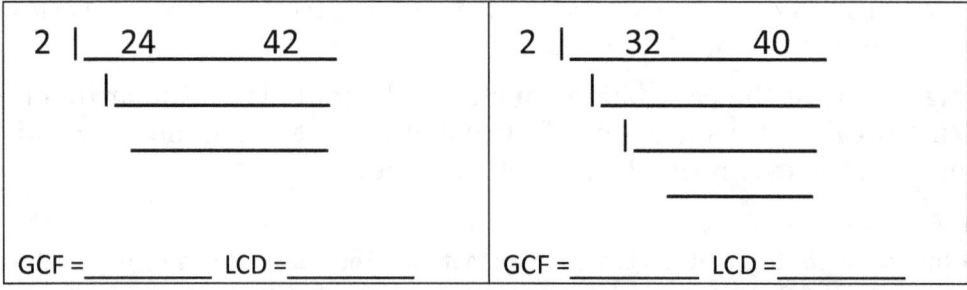

```
 2 |  24        42                 2 |  32        40
    |_____                         |_____
    _____                          |_____

    _____                          _____


 GCF =_____  LCD =_____      GCF =_____  LCD =_____
```

- Answers are on the next page –

ANSWERS: (Exercise 14)   GCF and LCD using *Upside-down birthday cake* method

24 & 42 (GCF = 6; LCD = 168)   32 & 40 (GCF = 8; LCD = 160)

```
        2 | 24      42                    2 |  32      40
        3 | 12      21                    2 |  16      20
      (2x3=6)   4       7   (4x7=28)      2 |   8      10
              (6 x 28 = 168)         (2x2x2=8)   4       5      (4x5=20)
                                            (8 x 20 = 160)
  GCF =  6     LCD =  168             GCF =  8     LCD =  160
```

Venn diagram method for finding GCF and LCD

The upside-down birthday cake method is easy.   Yet some people find it frustrating to work with. Another method involves using a Venn diagram.

> *Note*:
>
> A Venn diagram, or set diagram, was first created by John Venn in 1880. John Venn called them "Eulerian Circles" and used them to visually represent logical constructs (known as symbolic logic.) They are made up of a series (two or more) of circles that overlap. Each circle represents one set of values for a specific object. Once constructed, each circle has all the values associated with that object. If each object has common values, they are placed in the area where the circles overlap (in the center.)

To solve for GCF and LCD using Venn diagrams, you will use the *prime birthday cake factoring* method that was previously covered in this chapter.

Solving for GCF using Venn diagrams

Before focusing on Venn diagrams to determine the GCF of two numbers, a brief review of the *prime birthday cake factoring* method is necessary. To determine the GCF of the two numbers 12 and 18, you need to determine all of the prime factors of each number. Using the *prime birthday cake method*, you would follow these steps:

1.   Find the prime factors for the number 12 using the *prime birthday cake method*
2.   Find the prime factors for the number 18 using the *prime birthday cake method*

Here are the two birthday cakes for the numbers 12, and 18 – remember to always work the problem until you have the candle (the number 1) on top:

```
         1                          1
       3 | 3                      3 | 3
       2 | 6                      3 | 9
       2 |  12                    2 |  18
```

Using the birthday cake method, you see that the prime factors for the number *12* are *2, 2,* and *3*. For *18* they are *2, 3,* and *3*.

Once you have determined the prime factors, using the *birthday cake* method, for each number, you are ready to use them in a Venn diagram.

First you need to draw two circles that overlap each other – a Venn diagram. Initially it will have no numbers in it. Title each circle of the diagram by placing a 12 above the left circle and an 18 above the right one. (See the diagram below.)

Follow these continuing steps to determine the GCF using a Venn diagram:

3. Look at the prime factors for each number and find the common numbers in both.
   12 (2, 2, 3) and 18 (2, 3, 3)   BOTH have a 2 and 3
4. Write those numbers (2 and 3) in the intersecting region of the Venn diagram.
5. Multiply all values in the intersecting region of the diagram together to obtain the GCF. You can put the answer in the intersection region or below the Venn diagram.

At this point, your Venn diagram should look like this:

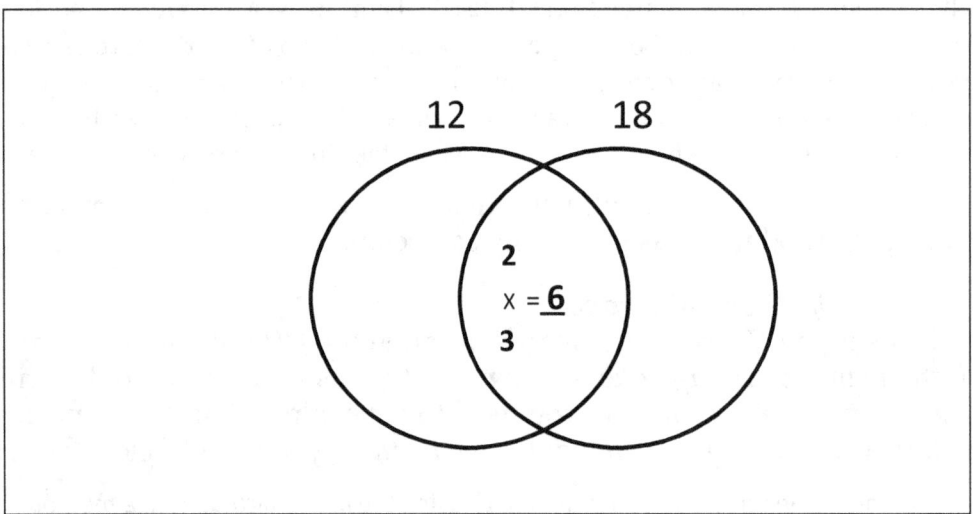

The greatest common factor for the number 12 and 18 is 6. Just multiply 2 x 3.

When using a Venn diagram, you place all prime factors that are the same for both numbers, in the common area (intersecting) of the circles and then multiply those values together to get the GCF.

Solving for LCD/LCM using Venn diagrams

Once you have determined the GCF of the two numbers using the Venn diagram you can continue to use the same diagram to determine the LCD/LCM. To quickly review, you got to this point by doing the following:

- Created birthday cakes to find the prime factors for each number (12 and 18)
- Used the Venn diagram to find the GCF

The actual work should look similar to this –

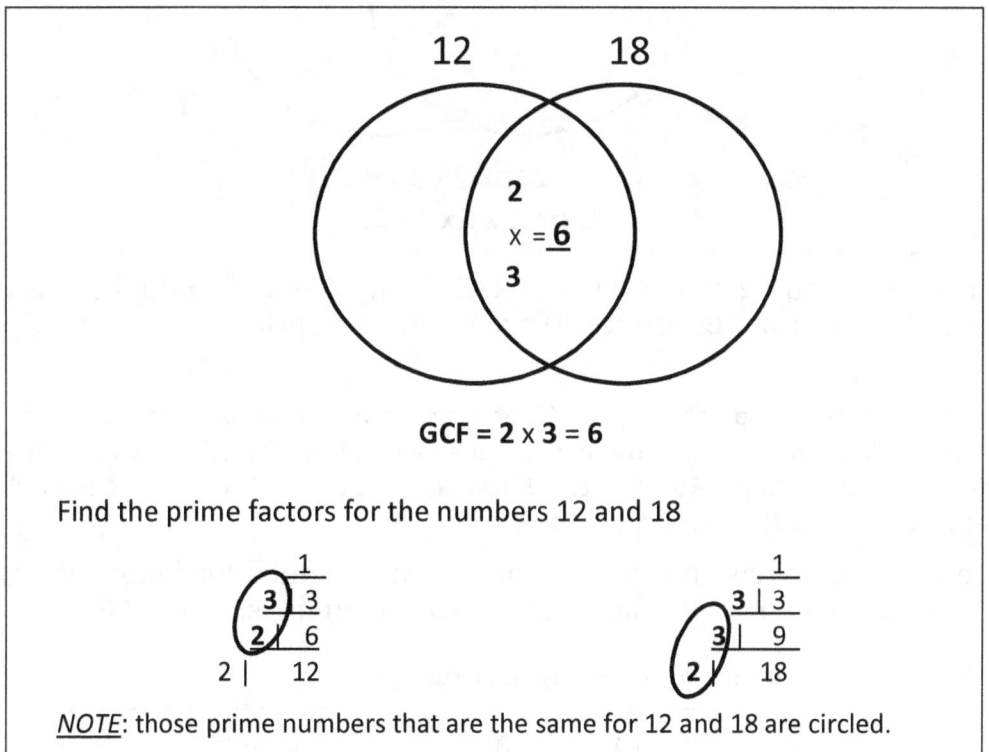

GCF = 2 x 3 = 6

Find the prime factors for the numbers 12 and 18

NOTE: those prime numbers that are the same for 12 and 18 are circled.

Looking at your work for finding all prime factors of both numbers and the Venn diagram, you are ready to determine the LCM. After you place all like prime factors for each number in the center of the diagram, look for any prime factors for each number that are not common. In this case, the number *12* has an extra *2* and the number *18* has another *3*. Take the unique *2* from the primes for the *12* and place it in the open area of the *12* circle. Do the same with the *3*, placing it in the *18* circle.

Once you have placed the non-common numbers in each circle you are ready to determine the LCD for them. Take the *2* in the *12* circle and multiply it by the GCF in the intersection area (*6*); giving you a product of *12* and then multiply that product times *3* (in the *18* circle) to get the final LCD of *36*.

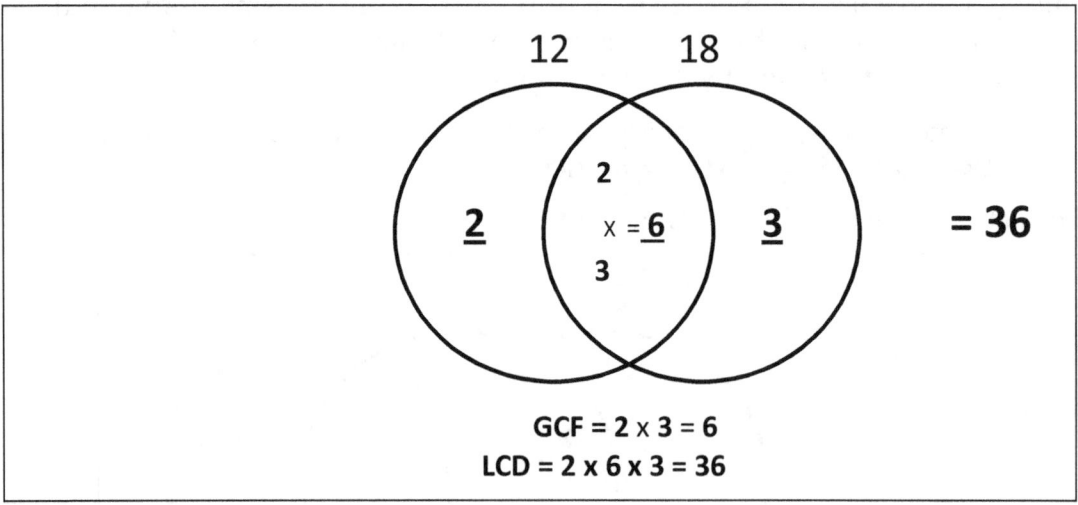

GCF = 2 x 3 = 6
LCD = 2 x 6 x 3 = 36

The Venn diagram above shows the final LCD is the result of multiplying *2* (the remaining prime factor from *12*) *x 6* (GCF) x *3*! (Remaining prime factor from *18*)

Note:
There is another way to determine the LCD once the Venn diagram is completed. If you look at the diagram, you can cross multiply the value (*3*) in the *18* circle times the original value of *12*, to get *36*. You could have also multiplied 2 (from the 12 circle) times *18* to get *36*. *36* is the LCD for 12 and 18.

It is time to do an exercise using the Venn diagram method for finding the GCF and LCD. Find the GCF / LCD for the following sets of numbers: (20 and 8).

EXERCISE 15 (remember to use the Prime Birthday cake factoring!)

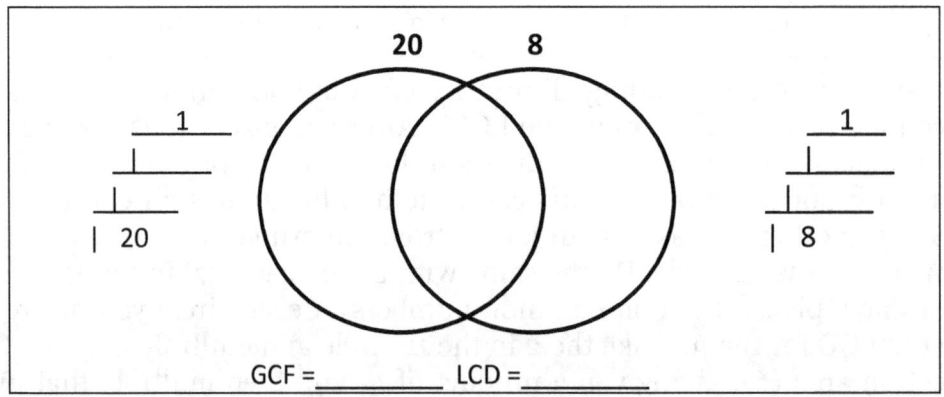

— Answers are on the next page —

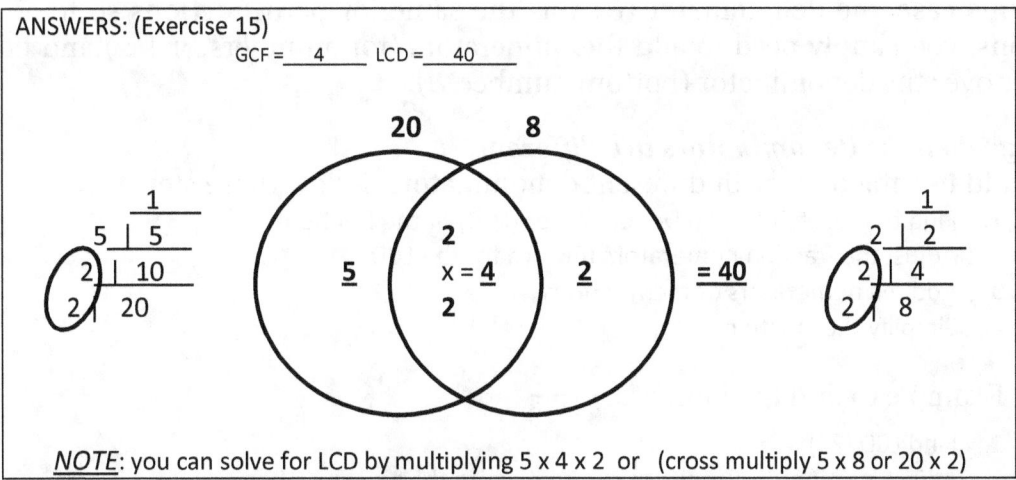

ANSWERS: (Exercise 15)

GCF = 4   LCD = 40

NOTE: you can solve for LCD by multiplying 5 x 4 x 2 or (cross multiply 5 x 8 or 20 x 2)

With a full understanding of the two terms *Least Common Denominator* (LCD) and the *Greatest Common Factor* (GCF) and how to solve for them; it is time to move onto fractions and their math.

What are Fractions?

Fractions are made up of two numbers; one over another. The top number is called the numerator; the bottom is known as the denominator. Officially it is a rational number expressed in one of two forms.

a/b – in-line notation,   or $\dfrac{a}{b}$ – traditional notation where **a** and **b** are integers

$\dfrac{\text{Numerator}}{\text{Denominator}}$ $\dfrac{5}{8}$	Numerator/Denominator **5/8**

Adding Fractions

Adding fractions is easy; but, the denominators (bottom numbers) must be the same to add them. If they aren't the same, you need to find a way to create a denominator that both fractions can use. You will need to find the *Least Common Denominator* (LCD).

Adding when the Denominators are the same

To add two fractions with the same denominator (bottom number), add the numerators (top numbers) and place that sum over the common denominator:

$$\frac{3}{8} + \frac{2}{8} = \frac{5}{8}$$

In this case, the denominator (8) was the same for both fractions so to add the fractions, you simply need to add the numerators (top numbers, 3 + 2) and put the answer over the denominator (bottom number, 8).

Adding when the Denominators are different
To add two fractions with different denominators, follow these steps:
1. Find the Least Common Denominator (LCD) of the fractions
2. Adjust the fraction numerators upward for the LCD
3. Add the numerators of the fractions
4. Simplify the Fraction

For Example to find the Sum of $\dfrac{2}{3} + \dfrac{3}{8}$
1. Find LCD (24)
2. Adjust fractions to use the LCD – 24 (upsize their top numbers)
3. Add the numerators (top numbers) of fractions (16 + 9 [25])
4. Simplify the fraction if possible

$$\frac{2}{3} + \frac{3}{8} \gg {}^{16}/_{24} + {}^{9}/_{24} \gg {}^{25}/_{24} = 1\frac{1}{24}$$

<u>Note</u>:
When adjusting the top numbers for the LCD, if you used the Venn diagram method to determine the LCD, you can cross multiply the remaining values in each circle times the top number. For example if you create a Venn diagram for the denominators 3 and 8 you will find that the GCF is 1 and the LCD is 24 – with 3 in the 3 circle and 8 in the 8 circle. To upsize 2/3 to x/24 just multiply 2 x 8 (16) / 24; and 3/8 becomes 3 x 3 (9)/24.

Subtracting fractions

The method of subtracting two fractions is similar to the process of adding fractions. It all depends upon the denominator's value – the same or different.

Subtracting when the Denominators are the same
To subtract two fractions with the same denominator, subtract the numerators and place that sum over the common denominator.

$$^8/_{15} - {}^2/_{15} = {}^6/_{15} \; which \; simplifies \; to \; {}^2/_5$$

In this case, the denominator (15) was the same for both fractions so to subtract the fractions, you simply need to subtract the numerators (top numbers: 8 – 2) and put the answer over the denominator (15). 6/15 can be further simplified to 2/5 by dividing each part by 3 – so the answer is 2/5.

Subtracting when the Denominators are different

To subtract two fractions with different denominators, follow these steps:

1. Find the Least Common Denominator (LCD) for both fractions
2. Adjust the fraction numerators upward for the LCD
3. Subtract the numerators of the fractions
4. Simplify the Fraction

For Example to find the Difference of 2/3 – 3/8

1. Find LCD by multiplying the denominators (3 * 8 = 24)
2. Adjust fractions to use the LCD – 24
3. Subtract the numerators of fractions (16 – 9 = 7)
4. Simplify the fraction if possible.

$$\frac{2}{3} - \frac{3}{8} \gg \frac{16}{24} - \frac{9}{24} = \frac{7}{24}$$

Multiplying with Fractions

Unlike addition and subtraction, when multiplying fractions you do not need to find a common denominator before doing the math.

When multiplying two fractions the result will be a fraction with a numerator that is the product of the fraction's numerators and a denominator that is the product of the fraction's denominators.

For example multiply

$$2/3 \times 3/8 = {}^{2 \times 3}/_{3 \times 8} = {}^{6}/_{24} = {}^{1}/_{4}$$

Notice that we multiply 2 x 3 for 6 on the top (numerator) and 3 x 8 for 24 on the bottom (denominator) and then we simplified the answer to 1/4.

You can also multiply whole numbers with fractions in a similar way. To multiply 4 times 3/8 just turn the whole number 4 into a fraction and do the multiplication –

$$4 \times {}^{3}/_{8} \gg {}^{4}/_{1} \times {}^{3}/_{8} = {}^{12}/_{8} = 1\,{}^{1}/_{2}$$

So multiplying by fractions is a breeze! Just multiply the top by the top and the bottom by the bottom and simplify if necessary.

Dividing with Fractions

Dividing by fractions is just like multiplying by fractions, except that you need to perform one additional step! First you need to find what is known as the *reciprocal* of the dividing fraction (divisor number). That is the fraction that is the divisor (the number being divided into another number.)

Note: RECIPROCALS

A reciprocal of a fraction is the inverse of the fraction. In other words, just flip it over — what was on top is now on the bottom and what was on the bottom is now on the top. For instance:

The reciprocal of $\frac{1}{5}$ *is* $\frac{5}{1}$; $\frac{3}{8}$ *is* $\frac{8}{3}$ and so on ...

Tip:

To divide by a fraction, simply "flip over" the fraction and then change the sign to a multiply and multiply the two numbers.

To divide any number by a fraction you will need to find the divisor fraction's reciprocal and then convert the problem to a multiplication problem. This simply means that you need to "flip" the fraction in the divisor part of the problem and multiply it by the top part of the fraction — the problem becomes a fraction multiplication problem.

Here is an example problem:

$\frac{4}{3/8}$ To solve this problem you need to first determine the reciprocal for 3/8.

Once you determine the reciprocal — 8/3, convert 4 to the fraction 4/1 and then multiply the two numbers together! So the problem should look similar to this:

$$\frac{4}{3/8} = \frac{4}{1} \times \frac{8}{3} = \frac{32}{3} = 10\frac{2}{3} \text{ so 4 divided by 3/8 equals 10 2/3!}$$

One final example involves two fractions, 2/3 divided by 3/4. Again all you have to do, to divide a fraction by a fraction, is inverse (flip over) the bottom fraction and then multiply. Here is the problem:

$$\frac{2/3}{3/4} = \frac{2}{3} \times \frac{4}{3} = \frac{8}{9} \text{ Thus 2/3 divided by 3/4 is 8/9!}$$

That is all there is to dividing by a fraction. Easy stuff!

Calculator Tricks (Part 3)

Using the calculator we will continue with those magical calculation tricks!

Good Luck or Bad Luck

1. Select a three-digit number and enter it twice into the calculator. (example: 123123)
2. State that the number is divisible by 11 – verify it by dividing by 11.
3. Announce that the result is also divisible by 13 – verify it by dividing by 13.
4. Now divide the answer by the original three-digit number.
5. Announce that the final answer is ....

Answer: 7! – NOTE: 11 x 13 x 7 = 1001 and 1001 times any three digit number is the number repeated twice 1001 x 321 = 321,321

Secret of 73

1. Select any four-digit number and enter it twice into a calculator. (example: 12341234)
2. State the number is evenly divisible by 137 – verify it on the calculator.
3. Now divide the answer by the original four-digit number.
4. Announce that the final answer is ...

Answer: 73! NOTE: 137 x 73 = 10001 and 10001 times any four digit number is the number repeated twice 10001 x 4321 = 4321,4321

Phone Number Trick

Remember to *PRESS [ENTER] AFTER EVERY STEP*

1. Enter the first three numbers of your phone number (NOT area code)
2. Multiply by 80   Press [=]
3. Add 1   Press [=]
4. Multiply by 250   Press [=]
5. Add the last four digits of your number   Press [=]
6. Add the last four digits of your number (AGAIN) Press [=]
7. Subtract 250   Press [=]
8. Divide by 2   Press [=]
9. Announce that the display shows ...

Answer: Their phone number in correct order! – NOTE: if you work this out using x for first three digits and y for last four, the formula works out to: $((250(80(x) + 1) + y + y) - 250)/2$. If you expand this you will get the following: $(20000x + 250 + 2y - 250) / 2$ or finally, $10000x + y$!

Math is Logic – or – The Logical Methods of Math (Part 2)

Logic and logic problems have been around since early mankind. Interestingly the logic of reasoning and its principles developed in different cultures – China, India, and Greece. Today's logic has its root in Greek Aristotelian logic.

The Greek thinker, Aristotle (384-322 BC/BCE), and followers, the *Peripatetics*, actively practiced the same logic as today. Aristotle wrote six works on logic; his followers referred to them as the *Organon* (Greek word meaning "tool.")

Of course Aristotle was influenced by the teachings handed down over the years; probably starting with the precursor works of the Greek philosopher Parmenides (5th Century). Other definitive lines of great Greek philosophers continued to refine the understanding of logic: Socrates (470 – 399 BC/BCE) and Plato (428 – 348 BC/BCE). Plato was Aristotle's teacher and mentor.

We continued to use the same process of logic in mathematics today. Here are a few problems to ponder concerning the process of logic. See if you can solve them:

Puzzle 1: Crates of Fruit

You are on an island and there are three crates of fruit that have washed up in front of you. One crate contains *only* apples. One crate contains *only* oranges. The other crate contains *both* apples and oranges. Each crate is labeled. One reads "apples", one reads "oranges", and one reads "apples and oranges". You know that NONE of the crates have been labeled correctly - they are all wrong.

If you can only take out and look at just one of the pieces of fruit from just one of the crates, how can you label ALL of the crates correctly?

Answer: Remove one fruit from the crate labeled "apples and oranges". (1) If it is an apple, then this crate contains "APPLES" only. This means that the crate labeled "apples" contains "ORANGES". Finally, this leaves the crate labeled oranges which contains "APPLES and ORANGES". (2) If you remove an orange, then the crate contains "ORANGES" only. This means that the crate labeled oranges contains "APPLES" and the apples crate contains "APPLES and ORANGES".

Puzzle 2: Matching Socks

You are in a rush and in the dark you reach into your sock draw to get a pair of matching socks. If your sock drawer has 6 black socks, 4 brown socks, 8 white socks, and 2 tan socks, how many socks would you have to pull out in the dark to be sure you had a matching pair?

Answer: 5! There are only four different colors of socks. Assuming you pull out one of each on the first four draws, the next draw will have to be one of the colors.

Puzzle 3: Relative Picture

A man on a park bench is looking at a small portrait. You ask him, "Who is that in the picture?" The man says, "Brothers and sisters, I have none, but that man's father, is my father's son."

Can you tell who the person is in the picture?

Answer: The picture is of his son!

Puzzle 4: The Boy and Girl

A boy and a girl are talking. "I am a boy" - said the child with black hair. "I am a girl" - said the child with red hair.

At least one of them lied. Who is the boy and who is the girl?

Answer: The black-haired child is a girl, and the red-haired child is a boy. The LOGIC: If at least one is lying, *and* there is one of each sex on the bench, then *both* must be lying. If only one was lying, then there would be two children of the same sex. Since the latter would not be following the given rules, then it is concluded that each child is of the opposite sex that they say they are.

Puzzle 5: My Sons

I have two sons, Richard and David. David is the youngest. Together their ages add up to 20. In five years, Richard will be twice as old as David. How old are Richard and David today?

Answer: 15 and 5: This is actually a simple problem involving two algebraic problems and substitution. Assume x is Richard's age and y is David's age. So x + y = 20. Then in five years (x+5) = 2(y+5) or x = 2y + 5. Now substitute (2y+5) for x in the first problem – (2y +5) + y = 20 or 3y = 15 or y = 5. So today David is 5 and Richard is 15 (20 – 5).

Puzzle 6: Twin Brothers

This is a similar problem to the one given in the last logic section (a fork in the road to the town of truth or lies).

There are twin brothers; one which always tells the truth and one which always lies. What single yes/no question could you ask of either brother to figure out which brother is which?

Answer: Would your brother say that you tell the truth? Why? The key is to find a question where the lying brother will give an opposite answer from his brother. The lying bother will answer this with "Yes"; while the truth brother will answer "no".

Were you able to solve these problems correctly? If you answer YES, then you are a logic genius!

Simple Math is All You Need (Part 3)

Today you will learn other rapid ways of multiplication. These methods include multiplying any number times 12; multiplying any two numbers less than and near 100, squaring numbers ending in 0 or 5 and a couple more rapid math methods.

Multiplying any number times 12

On Day One you learned how to quickly multiply any number by 11; even very large numbers. If you recall you wrote the number down, drew a line below it, added a leading zero, and then wrote the answer immediately below it. The process you used was to start at the right most position of the number and add its neighbor (to the right), writing down the value and carrying if necessary. Then you repeated this process until you had the answer written below the number to be multiplied.

You can quickly multiply any number by 12 by following a similar series of steps:

1. Write a leading 0 before the number and draw a line below it for the answer.
2. Multiply the first digit by 2 and write it down (it has no neighbor to add to it).
3. Multiply the next digit by 2 and add its neighbor (to the right) to the sum and write it down.
   - Remember if the number is greater than 9 add the tens to the next number
4. Repeat step 3 until you reach the end of the number.
   - Don't forget to add the last number to the 0 digit and write it down.

Now use the problem: *987234 x 12* and solve it following these steps

(the actual problem is shown below)

1. Double 4 and add nothing (no neighbor), write 8
2. Double 3 (6) + 4 (neighbor) = 10, write 0; carry 1
3. Double 2 (4) + 3 + 1 (carry) = 8; write 8
4. Double 7 (14) + 2 = 16; write 6; carry 1
5. Double 8 (16) + 7 + 1 (carry) = 24; write 4, carry 2
6. Double 9 (18) + 8 + 2 (carry) = 28; write 8; carry 2
7. Double 0 (0) + 9 + 2 (carry) = 11; write 11 (last digit)

$$2\ 2\ 1 \quad 1 \quad \curvearrowright\curvearrowright$$
$$\underline{0987234}$$
$$11846808$$

And that is all there is too it!

This system is credited to the late Professor Jakow Trachtenberg, founder of the Mathematical Institute in Zurich, Switzerland (1950). He came up with this and other mental math manipulation in WWII while a prisoner in a Nazi concentration camp.

This same method is one of the Indian Vedic math methods of multiplication. It can be found in Bharati Krishna Tirthaji's book, *Vedic Mathematics* (1965), which is the starting point for all work on Vedic math. He began lecturing about this discovery in 1958. Vedic comes from the Sanskrit word "Veda" meaning knowledge.

> *Note*:
> Try the same method multiplying by 13 or 14 or 15 ... up through 19 !
> Remember to multiply each digit by the unit number (2, 3, 4, etc.) and then add the neighbor.

Multiplying two numbers less than and near 100

If you have two numbers that are close to 100, and less than 100, you can quickly multiply them together following these steps:

STEPS:

1. Make each number go to 100 (what number needs to be added to reach 100?)
   (96 needs 4 – 99 needs 1)
   PLACE their numbers above each number.

2. Cross subtract to verify the same number from each subtraction. You can do this below the numbers.

   This becomes the left two digits (first part) of the four digit answer

3. Multiply the difference between the two numbers (from 100) 4 x 1 = 4
4. Place a leading 0 in front of the answer if it is less than 10 (04)

   This becomes the right two digits (last part) of the four digit solution

PROBLEM 1:

$$\overset{+4}{96} \times \overset{+1}{99} = 95 \quad 04$$

– 1 = 95   – 4 = 95   < = cross subtract to obtain the left part of answer

4 x 1 = 4   < = multiply the two differences from 100 to obtain right part of answer

Answer is 95 04

PROBLEM 2:

$$\overset{+7}{93} \times \overset{+8}{92} = 85 \ 56$$

– 8 = 85   – 7 = 85

7 x 8 = 56

Answer is 85 56

EXERCISE 16

+ __ + ____ 88 x 96 = ___ ___ - __ - ____	+ __ + ____ 97 x 89 = ___ ___ - __ - ____

- Answers are on the next page –

ANSWERS: (Exercise 16)
(1) <u>8448</u>   (2) <u>86 33</u>

$$+ \underline{12} \quad + \underline{4}$$
$$88 \times 96 = \underline{84\ 48}$$
$$\underline{-4} = 84 \quad \underline{-12} = 84$$

$$+ \underline{3} \quad + \underline{11}$$
$$97 \times 89 = \underline{86\ 33}$$
$$\underline{-11} = 86 \quad \underline{-3} = 86$$

Squaring any number ending in 0

RULE:
Square digits before final 0 and add two zeros

$$30^2 = 30 \times 30 = (3 \times 3)\ \underline{00} = = > 900$$

$$60^2 = 3600$$

$$70^2 = 4900$$

$$120^2 = 14400$$

That's all there is to it! Just add two zeros if there is one zero. Of course if you are squaring a large number with two zeros (like 900, you double the 0s to four from two – so 900 x 900 is 810000!

Squaring any number ending in 5

RULE:
Multiply digits before final 5 by next HIGHER number and append 25 to the end

$$35^2 = 35 \times 35 = (3 \times 4)\underline{25} \qquad = = > 1225$$

$$65^2 = 65 \times 65 = (6 \times 7)\underline{25} \qquad = = > 4225$$

$$55^2 = 55 \times 55 = (5 \times 6)\underline{25} \qquad = = > 3025$$

$$115^2 \qquad = (11 \times 12)\underline{25} \quad = = > 13225$$

The 25 at the end is the square of 5 (5 x 5). This is a powerful way to square any number that ends in 5. This and the previous short-cut will be used later when finding _any_ number's square.

Another rapid method for multiplying any number less than 100 by 99

You can quickly multiply any number less than 100 by 99 by following these steps:

1. Subtract 1 from the number to be multiplied by 99
2. Subtract the number to be multiplied by 99 from 100
3. Combine the two results into an answer
   – Step 1 is the first half and Step 2 the second half of the answer

Here are a couple of examples:

$22 \times 99 = $ _____ $35 \times 99 = $ _____
$22 - 1 = 21$ $35 - 1 = 34$
$100 - 22 = 78$ $100 - 35 = 65$
$= 2178$ $= 3465$

It really is that easy!
Now you try to multiply the following two problems:

EXERCISE 17

$45 \times 99 = $ _____	$78 \times 99 = $ _____
____ $- \underline{1} = $ ____	____ $- \underline{1} = $ ____
$100 - $ __ $ = $ ____	$100 - $ __ $ = $ ____

– Answers are on the next page –

Hope you found this method a powerful, easy way to multiply by 99!

Rapid Multiply Two digits times 101

Do you ever need to multiply a two-digit number by 101 (24 x 101 or 53 x 101)? If so, this is a great method. Here is the trick – just write the two digit number down twice, one next to the other. For example:

$24 \times 101 = 2424$ $72 \times 101 = 7272$ $52 \times 101 = 5252$ $43 \times 101 = 4343$

Multiplying any two digit number by 100 is the same as writing the number and placing two zeros after it. If you multiply 1 by any number it equals the number.

Since multiplying two digits by 101 is the same as multiplying by 100 and then by 1 and adding the results you get the two digit number twice.

```
ANSWERS: (Exercise 17)
   (1) 4455                          (2) 7722

      45 x 99 = 4455                    78 x 99 = 7722

        45 – 1 = 44                      78 – 1 = 77
        100 - 45 = 55                    100 – 78 = 22
```

Rapid Multiply Three Digits by 1001

Following the same logic for two digit numbers multiplied by 101, you can take any three digit number times 1001 and write the answer automatically.

Just write the three digit number (e.g. 246 x 1001, 781 x 1001, and so on) twice. For example:

246 x 1001 = 246246   781 x 1001 = 781781   892 x 1001 = 892892

Again 1000 times any number is the number with three zeros. One times any number is the number. If you multiply a three digit number times 1000 and then times 1 and add the values together you get the three digit number twice.

Multiplying two teen numbers that differ by 1

There are times that you need to multiply two numbers between 10 and 19 that are only one digit off; for example 13 x 14 or perhaps 17 x 18. You could multiply them the traditional way by writing both numbers down and solving the problem.

If you know the square of one of the numbers you could start with it and then either add or subtract the same number from that result.

Yet there is a third way – and here are the steps:

– These – steps will demonstrate the process of multiplying 13 x 14 –
1. Take higher number and add the unit from the lower number (e.g. 13 x 14 = 14+3 = 17)
2. Take the result of that number multiply by 10 (e.g. 17x10 =170)
3. Multiply the unit numbers together and add that value to sum (e.g. 4x3=12 ... 170+12)

Here are a few examples to demonstrate the steps:

```
   15 x 16 = _____              17 x 18 = _____
     16 + 5 = 21                    18 + 7 = 25
     21 x 10 = 210                  25 x 10 = 250
     5 x 6 =+ 30                    8 x 7 =+ 56
          240      <<= = =  Answer  = = =>>    306
```

So there you have it, another short cut for multiplying two numbers.

Endgame for the Day

Today's problem requires your use of both logic and your ability to "think outside of the box!" So have fun thinking about the solution. Just remember, don't over think the problem!

If you are having trouble with this problem, create a model using pennies and four strings. Just layout the nine pennies (or any coins) into a 3 x 3 grid and make your strings longer than the diagonal points from dot to dot.

Connect the dots

Your challenge is to connect the nine dots laid out in a three by three matrix by drawing four straight lines without lifting your pencil from the paper:

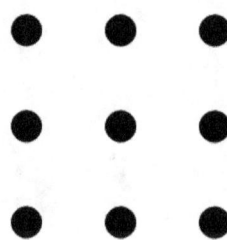

Remember:
Do not lift your pencil – you can draw 4 connecting lines to connect all the dots.

Finally, the only hint I will give you is to remember to *think outside of the box*!

– Answer is on the next page –

ANSWER: (Connect the Dots)

If you have already given up, the solution is found here along side of the explanation.

To draw four lines and not lift your pencil,

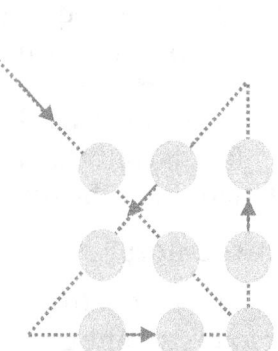

1. Start outside of the box of dots (top left
2. corner)
3. Draw the first line diagonally down through the first three dots.
4. Continue to draw up the right side through the next two dots and go beyond the dots.
5. Then turn downward and left covering the next two dots; and
6. Finally, draw to the right and connect the last two

Day Four

WELCOME TO DAY 4 OF MATH 4 2-DAY! The journey continues into the depths of mathematic reality.

Again, if you need to review any material that has been covered over the past three days, you should go back and review it. Today we will discuss some of the different ways people have performed multiplication over the years and gain an understanding of enormous numbers in the world of mathematics.

Today's Outline

Quotations of the Day
Bit of Fun in Math
    Math Concepts over the Years
Mathematical Curiosities – (Part 3)
    Enormous numbers and ways people visualize them
Math is Logic – or – The Logical Methods of Math (Part 3)
    Probabilities and What are the Odds
Curious methods for Multiplication
Calculator Tricks – (Part 4)
Simple Math is All You Need – (Part 4)
    Determine if a number is evenly divisible by 2 through 10
    Multiplying any same number of digits by the same number of 9's
    Summing a series of sequential numbers
Endgame for the Day
    Logic Puzzles for Fun – 10 balls in 5 rows & Cupid's Arrow

Objectives:

(1) Define and visualize large numbers and math
(2) Perform multiplication using different methods
(3) Continue to build upon your knowledge of logic
(4) Work with the concept of probability
(5) Continue to perform mental math – avoiding the calculator

Quotations of the Day

∞ · ∞

Mathematics is the sister, as well as the servant, of the arts and is touched with the same madness and genius!

-- Harold Marston Morse

Harold Marston Morse (1892 –1977) was an American mathematician best known for his work on the calculus of variations in the large; a subject where he introduced the technique of differential topology now known as Morse theory. In 1933 he was awarded the Bôcher Memorial Prize for his work in mathematical analysis.

∞ · ∞

Mathematics is a game played according to certain simple rules with meaningless marks on paper.

-- David Hilbert

David Hilbert (1862 –1943) was a German mathematician who is one of the most influential and universal mathematicians of the 19th/early 20th centuries. He invented or developed a broad range of fundamental ideas in invariant theory, the axiomatization of geometry (predictive logic), mathematical infrastructure for quantum mechanics and general relativity. He is one of the founders of proof theory, mathematical logic and the distinction between math and meta-mathematics. He submitted a list of 23 unsolved (open) problems at the International Congress of Mathematicians in Paris in 1900.

∞ · ∞

An educated mind is, as it were, composed of all the minds of preceding ages!

-- Bernard de Fontenelle

Bernard de Fontenelle (1657 - 1757) was a French mathematician who wrote on the history of mathematics and the philosophy of mathematics and science. Voltaire said he was the most universal mind in the era of Louis XIV. Many ideas of the Enlightenment are found in his works.

∞ · ∞

Bit of Fun in Math

This week's topic is an educational satire on the "*State of Math*" over the years.

Math over the Years

Word problems have always been challenging for many students. Add to this, issues like being politically correct yet being challenging for all levels of students. Here is a simple analogy of word problems over the ages for your amusement.

(It is believed that a similar chart was first published in Reader's Digest, February 1996 Issue)

1950s and 60s arithmetic test:	**1970s new-math test:**
"A logger cuts down and sells a truckload of lumber for $100. His cost of production is 4/5 of the price (or $80). What is his profit?"	"A logger exchanges a set of lumber ("L") for a set ("M") of money. The cardinality of set M is exactly 100. Each element is worth $1.00 USD The cost of production (set "C") contains 20 fewer points than set "M". Now the question: What is the cardinality of set "P" of profits?"
1980s "dummied-down" version:	**1990s version:**
"A logger sells a truckload of lumber on the side of the road for $100. *Her* cost of production is $80; making her profit $20. Find and circle the number 20."	"An ignorant logger, not knowing any better, cut down a beautiful forest of 100 trees. He made a $20 profit. What do you think about this way of making a living? Topic for discussion: What do you think about how the birds and forest animals feel?"

2000s "new math" version:

"A hard working logger spends an unknown amount of time cutting down a truckload of lumber. He takes it to the mill and sells it for $100. He figures that his cost for production is $80. Get your calculator out, Enter 100 minus 80 on your calculator. His profit is the answer – What is his profit and how did you arrive at it?"

Original Source: unknown original source – many similar lists on Internet titled same!

Mathematical Curiosities – (Part 3)

The topic this week is large numbers. Numbers, as pointed out previously are infinite; they have no limits or boundaries in space or time. This section discusses concepts of large numbers and how mathematicians have agreed to define them.

Enormous numbers and ways people visualize them.

A few days back, we spoke about large numbers when discussing one billion. We learned that the time it would take to count to 1,000,000,000 would take over 30 years ... 1,000,000,000/(60x60x24x365.25) = 31.68808781 years.

Although a billion is a large number, it is nothing in the world of mathematics. Many people hear numbers in the trillions with issues like the American National Debt being around 14.38 trillion ($14, 384, 078, 225, 097 @ 8:00 am, EST 18 May 2011) and U.S. Federal Government spending near $3.55 trillion. If it takes 31.69 years to count to a billion, it will take about 31,690 years to count to a trillion! How about more than 443,660 years to count to 14 trillion!

> *Note*:
> You can see a real time U. S. Treasury clock by visiting: http://www.usdebtclock.org/

Then there are the next highest numbers in the quadrillion or quintillion ranges and yet they are small numbers in the world of mathematics.

Large Numbers since the time of Archimedes

Large numbers have been around since the time of Archimedes (287 – 212 BC/BCE). He is best known for his principle/law of "upthrust" (buoyancy). The story goes, (can be found at numerous websites on line):

> ... According to Marcus Vitruvius [writer, architect, engineer], King Hieron had a new crown made in the shape of a laurel wreath. Archimedes was asked to see if it was made of solid gold or if silver was added by a dishonest goldsmith. The condition was that Archimedes had to solve the problem without damaging the crown. Therefore, he could not melt it down in order to measure its volume. The story continues with Archimedes taking a bath. While bathing, he noticed the level of the water rose as he got in. He realized that this effect could be used to determine the volume of the crown, and therefore its density after weighing it. He decided that the density of the crown would be lower if cheaper and lighter metals had been added.

After making this discovery, he jumped from the tub and ran through the streets yelling "Eureka! Eureka!" ("I found it!"). He was so elated that he forgot to dress.

Archimedes then formulated his principle of buoyancy which states,

> *The weight of fluid that a submerged object displaces is the same amount of buoyancy force applied to the submerged object.*

Archimedes is known for more than his principle of buoyancy. He created the Archimedes screw; a machine with a revolving screw shaped blade inside a tube like cylinder. It was used to drain ships loaded with liquid and to transfer water from a low-lying body of water into irrigation canals. Versions of the Archimedes screw are still in use today. He defined the laws of the lever; although he did not invent the lever. He also designed block and tackle pulley systems and invented or improved many weapons of war; including the "Claw of Archimedes" ("Ship Shaker"), the odometer, and improvements to the catapult.

He has been quoted, "Give me a place to stand on, and I will move the Earth!"

Yet, none of these could compare to his greatest works in the field of mathematics. Some mathematical proofs involve the use of *infinitesimals* in a similar way to modern integral calculus (long before Newton and Leibniz).

Archimedes answered problems with a high degree of accuracy; specifying limits where the answer existed. The technique is known as the *method of exhaustion*, and is employed to approximately the value of π (Pi).

One of his most famous mathematical works is *The Sand Reckoner*. He wanted to calculate the number of grains of sand that the universe could contain. He did this to challenge the notion that the number of grains of sand was too large to be counted. He created a name for his own big number – the *myriad;* one of the earliest big numbers identified. More on his Sand Reckoner later.

Using scientific notation to handle large and small numbers

The problem, when describing or working with big numbers, is defining how they will be represented. Imagine writing a sextillion (1 with 21 zeros behind it) each time when calculating a math problem. You will use a lot of paper to write the number – before you even start to do any operation on the number.

This is where Scientific notation comes in. Scientific notation was created to handle a wide range of values which occur in scientific study. 1.0×10^9, for example, means one billion, or a 1 followed by nine zeros: 1 000 000 000, and 1.0×10^{-9} means one billionth, or 0.000 000 001. Writing 10^9 instead of nine zeros saves the writers effort and problems when counting a long series of zeros to see the number.

A 100 (1 followed by 2 zeros) is 10x10; 10,000 is 10x10x10x10. Using scientific notation, the exponent increases by one, 10^1 (10), 10^2 (100), or 10^4 (10,000).

Archimedes and the Sand Reckoner

One of the earliest examples of working with large numbers is found in his book *The Sand Reckoner*. Archimedes defined a system for naming large numbers. One reason was to explain, in understandable terms, the solution to the problem of how many grains of sand are on the earth. He called numbers up to a *myriad* (10^8) "first numbers" and called 10^8, the "unit of the second numbers". Multiples of this unit then became the second numbers, up to this unit taken a myriad number of times, $10^8 \cdot 10^8 = 10^{16}$. This, 10^{16}, is known as "unit of the third numbers", whose multiples were the third numbers, and so on. Archimedes continued naming numbers in this way up to a myriad times the unit of the $10^{8\text{th}}$ numbers:

He then embedded $(10^8)^{(10^8)} = 10^{8 \cdot 10^8}$ this construction within another copy of itself to produce names for numbers up to $(10^8)^{(10^8)^{(10^8)}} = 10^{8 \cdot 10^{16}}$

Archimedes then estimated the number of grains of sand that would be required to fill the known Universe, and found that it was no more than "one thousand myriad of the eighth numbers" (10^{63}). This is a whole lot more than the sand on Earth.

Remember 1 billion is 10^9, so 10^{63} is a very large number. In fact the formal name is a *vigintillion*.

Who first came up with the term Scientific Notation?

Interestingly, the concept of Scientific Notation was first introduced and used by Archimedes when he wrote *The Sand Reckoner* and he introduced his big number, the myriad. He did not express this number in the form we know today, a raised power. However, he did work with this concept early on.

It wasn't until 1637, when, Rene Descartes, defined *x*, *y*, and *z* as values to represent unknown numbers and the use of *raised Arabic numbers* as indicators of powers.

Over time, Descartes concepts were embraced by the Mathematical community.

Scientific notation was known as exponential notation by scientists and mathematicians since the time of Descartes.

The actual term *scientific notation* has an unclear history. According to the publishers of the Oxford English Dictionary, who track the history of words and their introduction into society, the term first appeared in the third edition of the *New International Dictionary of the English Language* in 1961. It was not in the second edition published in 1934. It was used extensively in the late 1960s and early 1970s especially in computer science and when using scientific calculators.

Large Numbers by 10 to the power of:

Understanding large numbers by powers of ten is easy; just remember that each major name changes in powers of three (that is three number places). For instance, a thousand is 10^3 or more simply a 1 followed by three numbers (1,000, 1,999). So a million is 10^6 or a 1 followed by six numbers (1,000,000, 1,123,943, or 1,999,999).

10^6 Million
10^9 Billion
10^{12} Trillion
10^{15} Quadrillion
10^{18} Quintillion
10^{21} Sextillion
10^{24} Septillion
10^{27} Octillion
10^{30} Nonillion
10^{33} Decillion
10^{36} Undecillion
10^{39} Duodecillion
10^{42} Tredecillion
     and so it goes ...
10^{99} Duotrigintillion
10^{100} Google

> An easy way to find the value of the numbers in the scale to the left is to take the number indicated by the prefix (such as 2 in *bi*llion, 4 in *quadri*llion, 9 in *deci*llion, etc.), add one to it, and multiply that result by 3. For example, in a trillion, the prefix is *tri*, meaning 3. Adding 1 to it gives 4. Now multiplying 4 by 3 gives us 12, which is the power to which 10 is to be raised to express a trillion in scientific notation:
>
> one trillion = 10^{12}.

The Googol family

The names *googol* and *googolplex* were invented by Edward Kasner's nephew, Milton Sirotta, and introduced in Kasner and Newman's 1940 book, *Mathematics and the Imagination,* in the following passage:

Words of wisdom are spoken by children at least as often as by scientists. The name "googol" was invented by a child (Dr. Kasner's nine-year-old nephew) who was asked to think up a name for a very big number, namely 1 with a hundred zeroes after it. He was very certain that this number was not infinite, and therefore equally certain that it had to have a name. At the same time that he suggested "googol" he gave a name for a still larger number: "Googolplex". A googolplex is much larger than a googol, but is still finite, as the inventor of the name was quick to point out. It was first suggested that a googolplex should be 1, followed by writing zeros until you got tired. This is a description of what would actually happen if one actually tried to write a googolplex, but different people get tired at different times and it would never do to have Carnera a better mathematician than Dr. Einstein, simply because he had more endurance. The googolplex is, then, a specific finite number, with so many zeros after the 1 that the number of zeros is a googol.

Googol and Googolplex are 10 raised to a power:

VALUE	NAME
10^{100}	*Googol*
10^{googol} ($10^{10^{100}}$)	*Googolplex*

A Googol is probably nothing special. For instance; the total number of elementary particles in the known universe is about 10 to the power of 80. It is approximately equivalent to 70! (70 factorial is 70 x 69 x 68 x 67 x [....] x 3 x 2 x 1). This is a number that can easily be conceived of mentally.

In contrast a Googolplex's numerical dimension is outside the realm of physical reality. According to Carl Sagan, an astronomer and television personality in his book, *Cosmos: A Personal Voyage*,

> "Writing a googolplex in base-10 numerals (i.e., 1 followed by a googol of zeroes) would be physically impossible, since doing so would require more space than the known universe provides."

So Googolplex is a big number!

Other large numbers used in mathematics

There are many other large numbers used in mathematics. These include *Skewes'* number, *Avogadro's* number, *Graham's* number, *Moser's* number, and *Steinhaus – Moser* notation (used to conceptualize large numbers).

Many of these numbers and their names came about as a way to describe issues of length and time in the fields of physics and astronomy.

Skewes' number

Skewes' number refers to several extremely large numbers used by the South African mathematician Stanley Skewes.

In mathematics, this number is the smallest natural number x for which [$\pi(x) - Li(x) \geq 0$] and where $\pi(x)$ is the prime counting function and $Li(x)$ is the offset logarithmic integral. Now that you are completely lost, consider that there are two Skewes' numbers:

Skewes first number is approximately equal to –

$$10^{10^{10^{8.85 \times 10^{33}}}}$$

While his second number is approximately equal to –

$$10^{10^{10^{10^{1000}}}}$$

His numbers are used to describe the limits of one of the most famous unsolved mathematical problems – the Riemann hypothesis (1859).

Note:
Bernhard Riemann proposed a hypothesis in his 1859 paper titled, *On the Number of Primes Less Than a Given Magnitude*. The Riemann hypothesis concerns the distribution of prime numbers. He is known for the Riemann zeta function. His hypothesis is considered, by some mathematicians, to be the most important unresolved problem in pure mathematics. It is found in David Hilbert's list of 23 unsolved problems. Part of his hypothesis was proven in 1973 by Pierre Deligne when he proved that it held true over finite fields. The complete version of the hypothesis still remains unsolved today.

Avogadro number (constant)

The *Avogadro number* is named after an early nineteenth century Italian scientist – Amedeo Avogadro. Amedeo is credited with being the first person to realize that the volume of a gas is proportional to the number of atoms or molecules.

Avogadro number, also known as the Avogadro constant, uses the symbols: L and N_A, and is defined as the number of "entities" in one *mole*; or more specifically, the number of carbon-12 atoms in 12 grams of unbound carbon-12 in its rest-energy electronic state.

The current estimate in the pure mathematics world for this number is –

$$N_A = (6.0221415 \pm 0.0000010)10^{23} mole^{-1}$$

Note:
A mole is a SI (International System of Units) base unit that measures an amount of substance. The mole's use is usually limited to measurement of subatomic, atomic, and molecular structures.

Graham's number

Graham's number, named after Ronald Graham, is described as the largest number that has ever been *seriously* used in a mathematical proof. It is too large to be written in scientific notation; it needs special notation to write it down.

Graham's number is much larger than other well known large numbers such as a googol and a googolplex, and even larger than Moser's number, another well-known large number.

Graham's number uses a notation invented by Donald Knuth. The number is:

$$G = G_{63} \text{ in Knuth's notation.}$$

So Graham's number is considered the largest actual number that is used in a mathematical proof. It can be found in the Guinness Book of Records.

Knuth's Up-Arrow Notation

Donald Knuth published the following notation for huge numbers in 1976. Understand that the notation $[\,a \nearrow b \nearrow c\,]$ means $[\,a \nearrow (b \nearrow c)\,]$ etc.

1. $m \uparrow n$	$= m \cdot \ldots \cdot m$	(n terms) $= m^n$
2. $m \uparrow \uparrow n$	$= m \uparrow \ldots \uparrow m$	(n terms) $= {}^n[m]$
3. $m \uparrow \uparrow \uparrow n = m \uparrow \uparrow \ldots \uparrow \uparrow m$	(n terms) $= {}^n[m]^{\wedge\, n}[m]$	

To see *how fast the numbers grow* let us look at the first values for $m = n = 3$:

$3 \uparrow 3$	$=$	$3^3 = 27$
$3 \uparrow \uparrow 3$	$=$	$3 \uparrow (3 \uparrow 3) = 3 \uparrow 27 = 3^{27} = 7{,}625{,}597{,}484{,}987$
$3 \uparrow \uparrow \uparrow 3$	$=$	$3 \uparrow \uparrow (3 \uparrow \uparrow 3) = 3 \uparrow \uparrow 7{,}625{,}597{,}484{,}987 = [3]^{7{,}625{,}597{,}484{,}987}$ A power tower of 7,625,597,484,987 layers high
$3 \uparrow \uparrow \uparrow \uparrow 3$	$=$	$3 \uparrow \uparrow \uparrow (3 \uparrow \uparrow \uparrow 3)$ Even the rank of this number is unimaginable!

Using Knuth's up-arrow notation *Ronald Graham* derived the following huge number as an upper bound in a part of combinatorics called *Ramsey theory*. *Graham's number* is defined as:

0. Define $G_0 = 3 \uparrow \uparrow \uparrow \uparrow 3$.
1. Define $G_1 = 3 \uparrow \ldots \uparrow 3$ (G_0 up-arrows).
2. Define $G_2 = 3 \uparrow \ldots \uparrow 3$ (G_1 up-arrows).
    . . . etc..
63. 63. Define $G_{63} = 3 \uparrow \ldots \uparrow 3$ (G_{62} up-arrows).

Then $G = G_{63}$ is Graham's number.

Moser's number

Leo Moser created a number that could not be defined using scientific notation.

This number is known as Moser's number and is equal to '2 in a Megagon' which is equal to $2[2[5]] = 2[256[3]_{256}]$.

To describe his number he started by using a new notation system known as Stienhaus' polygon notation, which was created in 1983.

It uses triangles, circles, and squares to identify large numbers. He used naming conventions of *Mega* and *Megiston*. A mega is equivalent to a 2 in a circle - ② and a megistron is the number equivalent to a 10 in a circle - ⑩. Then there is a *Megagon* which is equivalent to a polygon with *mega* sides.

Moser extended Steinhaus' notation by continuing the sequence of polygons with a pentagon, an hexagon, and so on.

Here are some of the symbols defined in Stienhaus' polygon notation:

$\triangle{a}$ = (any number a in a triangle) means a^a

$\boxed{a}$ = (a number a in a square) is equivalent with "the number a inside a triangles, which are all nested."

⟨a⟩ = (a number a in a pentagon) is equivalent with "the number a inside a squares, which are all nested."

These are just a few of the different large numbers that exist in the world of pure mathematics.

There are many other really large numbers that you can learn about. These include notations like

- Conway-Wechsler Extension,
- Power Towers,
- Hyper-factorial and Super-factorial,
- Bowers' Array Notation,
- Steinhaus-Moser-Ackermann operators,
- Friedman sequences,
- Conway's Chained Arrow Notation,
- Bowers' Extended Operators,
- Lin-Rado Busy Beaver Function,

. . . and many more.

As you can see there are many really large numbers that have been defined in the field of mathematics. As you continue to work with numbers, there will come a time that you may become more familiar with these really large numbers of math.

Math is Logic – or – The Logical Methods of Math (Part 3)

In 1850, James Clerk Maxwell, a Scottish theoretical physicist and mathematician, defined logic as:

> "The actual science of logic is conversant at present only with things either certain, impossible, or entirely doubtful, none of which (fortunately) we have to reason on. Therefore the true logic for this world is the calculus of Probabilities, which takes account of the magnitude of the probability which is, or ought to be, in a reasonable man's mind."

He was not alone in his believing that the mathematics of probability is actually an extension of the generalization of Aristotelian logic.

Many scientists have written on this topic, especially during the last twenty-five years. These include notables like E. T. Jaynes (of Stanford fame), Phil Gregory and Tom Loredo, and D. S. Sivea (St. John, Oxford). All have made major contributions to the world of science as related to probability and engineering, data analysis, and other fields.

In fact, by the late 1980's the field of "*probabilistic logic*" has emerged. This field has made inroads into the studies of Artificial intelligence, Bioinformatics, Game theory, and the Philosophy of science.

Probability is

Probability is the mathematical method of representing the chance or likelihood of something (an event) happening. This "*probability of the occurrence of an event*" is expressed as a fraction or a decimal from 0 to 1. The potential of an event not occurring, or unlikely, will have a probability near 0, and events that are likely to happen have probabilities near 1. For example a coin has two sides – heads and tails (front/back). If you toss it in the air, what are the chances that it will 'come up' heads when it lands on a table?

Using probability we know that the probability is 1 in 2 or expressed as 1/2 (as a fraction) or 0.50 (as decimal) or 50% (as a percentage.) In fact every time you 'flip' the same coin the probability remains the same. Each flip is its own event and thus the probability remains constant. Taking this same concept of flipping a coin and putting it into practice, you should see that half of the time you toss the coin it will 'come up' heads and the remainder of the time it will be tails.

That is the theory and it can be borne out in practice; the more times you 'flip' a coin, the closer you will come to 50% heads – 50% tails.

EXERCISE 18

This is a great way to prove tossing a coin has a 50/50 chance of being heads. It is best if you have a few friends do the same exercise independently:

> Take a piece of paper (tally sheet) and draw a line down the middle of the paper (from top to bottom) and then label the left column HEADS the right one TAILS. Over the next week (5 days), take a penny and toss it 20 times a day and write down your results. Compare your results with your friends. Combine all of the results. The more times you, and your friends, try this experiment, the better chance you will be able to show that the probability ratio of 50/50 (heads / tails) is true.

History of Probability theory

Probability theory began in France, in 1654, with two French mathematicians, Blaise Pascal and Pierre de Fermat. They use to write each other about two specific problems found in games of chance.

The types of problems that Pascal and Fermat solved still influence the theory of probability. Today, probability theory is a well established branch of mathematics that finds applications in every area of scholarly activity from music to physics, and in daily life from predicting risks of new medicine to weather prediction.

In 1921, John Maynard Keynes wrote a book titled, *A Treatise on Probability*. This book was described in several newspapers of the time as "the definitive work on probability that has been published in the 20th Century." In the book he describes probability in this way:

> "Intuitively, the mathematical theory of probability deals with patterns that occur in random events. For the theory of probability the nature of randomness is inessential. According to the French mathematician Marquis de Laplace randomness is a perceived phenomenon explained by human ignorance. Other mathematicians in the late 19th/ early 20th century came to a realization that chaos has emerged as the result of deterministic processes."

Talk about adding confusion to a mathematical concept. Does this mean that the world around us is filled with chaos and randomness and we are simply ignorant?

Main Idea of Probability

The main idea of probability is actually easy to understand. It is that some event will occur and the question is, "What are the 'chances' of its occurring?"

More direct, the study of probability helps us figure out the likelihood of something happening. Simple! Direct! To the point!

Formula for the Probability of an Event

The exciting part of probability is the simple formula that is used to determine the 'probability' of something happening.

Probability (P) of an Event (E)
$$P(E) = \frac{Number\ of\ Ways\ an\ Event\ (E)\ can\ occur}{Total\ Number\ of\ Possible\ Outcomes}$$

P(E) in the formula is what we are trying to find out; what is the Probability of an Event occurring?

Working with probability is best demonstrated through examples.

Drawing a red marble from a bag

Image there is a cloth bag containing 4 red, 4 blue, and 2 black marbles. What is the probability of pulling a red marble out in one try? Obviously, any color marble can be removed from the bag on the first try.

Note:

In mathematical terms this means that, "all outcomes are equally likely: any individual marble has the same chance of being drawn."

The first thing to understanding is the meaning of "*Number of Ways an Event "E" can occur*" and "*Total Number of Possible Outcomes.*"

The *number of ways an event "E" can occur* has two parts: (1) The *event* is the chance of drawing a red marble from the bag. (2) The *number of ways it (E) can occur* is the possible number of ways we can draw out a red marble. There are four red marbles so the number of ways it can occur is 4.

The *total number of possible outcomes* is just the total "*sample space.*" The sample space consists of all the marbles in the bag. We are equally likely to pick any of the 10 marbles in the bag. 4 red + 4 blue + 2 black marbles gives us 10.

So, the *number of ways* a red marble (the *Event*) *can occur* is 4 and the total number of possible outcomes is the total number of marbles, 10. Plugging these numbers into the formula it will look like this:

$$P(E) = \frac{4}{10}\ or\ \frac{(Number\ of\ Ways\ and\ Event\ (E)\ can\ occur)}{(Total\ Number\ of\ Possible\ Outcomes)}$$

Using the formula for probability, the chances of drawing a red marble out of the bag on the first try is *4/10*, or *2/5* (fraction), 0.40 (decimal), or 40% as a percentage.

That was easy! We have a 40% or 2/5 chance of drawing out a red marble!

Probability of tossing three coins and all coming up 'heads'

For the purpose of this problem, you can use three pennies.

First we need to determine the *sample space*, or the number of possible outcomes. In this case, the *sample space* is the total number of combinations of heads and tails that can exist when throwing three coins.

Since each coin has two sides, HEADS and TAILS, to determine the total number of possible outcomes can be determined in this manner:

Each coin has two sides. One coin has two conditions or possibilities – coming up heads (H) or tails (T). Using two coins each coin has two conditions (H or T). There are 2 sides on each coin so 2 sides x 2 sides = 4 different events or conditions that can occur. These four combined sets of conditions when both coins are tossed are:

{HH} {HT} {TH} {TT}

Two coins have two conditions each; so the total number of events is 2 x 2 (or 4).

Following this logic, we can determine the total number of outcomes (or events) for three coins is 2 sides x 2 sides x 2 sides = 8 events. Here they are:

{HHH} {HHT} {HTH} {HTT} {THH} {THT} {TTH} {TTT}

Knowing the total number of outcomes, it is time to determine how many ways a particular condition or event can occur. In this case, the problem is "What is the probability of tossing three coins, one time, and they all come up heads?" Since each coin only has one head (H) and one tail (H) and the total number of events, as shown above is 8, there is only one state where all three coins come up heads. So the P(E) of 3 heads, or the probability of the event is 1 in 8 or 1 / 8 or 0.125 or 12.5%.

Take it another step – what is the probability of getting two heads and one tails? If you look at the coin combinations above you will see that the P(E) is –

3 / 8 – or – 0.375 – or – 37.5%

You can also figure it out using a little logic. If there are three coins and the condition you are looking for is two heads and one tail, just think about it. Since only one coin will be tails and the other two are heads and there are three coins – any one coin will have a tail – thus three different coins can have a tail while the other two are heads. Thus 3 of 8 total events.

Probability of a child's-gender in a two child family

For discussion purposes, assume that having a boy or girl is equally likely, even though statistically that may not be the case for a given population at a given time.

In a two-child family, there are four and only four possible combinations of children. We will label boys (B) and girls (G); in each case the first letter represents the oldest child:

$$\{BB\} \qquad \{BG\} \qquad \{GB\} \qquad \{GG\}$$

These are the only four combinations – thus any one combination has a 1-in-4 or 1/4 probability. Starting with this, it is time to move on to some problems.

> *Note*: In this example, the first incidence within the braces is the first born and the second event follows. For example {BG}, the first born is a boy and the second is a girl.

In a two-child family, one child is a boy. What is the probability that the other child is a girl?

This problem is simple to solve. Since one child is already a boy, the combination of {GG} is eliminated. Now the remaining three combinations are:

$$\{BB\} \qquad \{BG\} \qquad \{GB\}$$

Looking at the remaining three combinations, only 1 of the 3 has a *{BB}* combination, thus the chances of the 2nd child being a boy is 1-in-3 or 1/3. In contrast, the chances of the other child being a girl are 2-in-3 or 2/3rds.

Notice in the above problem, the order of birth and gender are not related. The next question deals with the gender and order of birth.

What if the older (first born) child is a boy? Does this information change the probability that the second child is a girl?

Knowing that the first born child is a boy eliminates two combinations both involving a girl as the first born – {GG} and {GB}. The first born must be a boy (B). So now there are only two combinations left:

$$\{BB\} \qquad \{BG\}$$

With only these two combinations, the probability that the second child will be a girl is 1-in-2 or 1/2.

This is only the beginning of the field of probabilities. There are many other areas to explore, including issues like: certain and impossible events; identify and solve for the complement of an event; mutually exclusive and non-mutually exclusive events; and dependent and independent events.

Curious methods for Multiplication

Today, you will hopefully be exposed to a few ways of performing multiplication that you weren't aware of. These include methods like *line crossing, lattice* or *gelosia multiplication, Gomutrika*, and *Br. Juan Diez's "Sumario Compendioso."* All of these methods offer alternative ways for performing multiplication.

You have already been introduced to another way of performing multiplication known as the *Trachtenberg method*. You used this method when you performed multiplication of large numbers by 11 and 12. It involved placing the number to be multiplied on a single line, drawing a line below it, adding a leading zero and writing the answer direct. If you recall you took the number, starting at the right, and added it to its neighbor (right) and wrote the result down. You repeated this step until you ran out of digits (including the leading zero).

History of multiplication

Methods of multiplication have been documented in the ancient civilizations of Egypt, Greece, China, the Indus Valley (India), and Babylon. Even the *Ishango* bone (a baboon bone tool found near the Nile River), which dates to 18,000 to 20,000 BC/BCE, suggests people had knowledge of multiplication by 2 during the Upper Paleolithic era of mankind.

> *Note*:
>
> The word, multiply comes from the roots *multi*, many, and *pli*, folds. It suggests that a number is folded on itself many times. Geoffrey Chaucer is credited with coining the phrase in his 1391 work, *A Treatise on the Astrolabe*.

India and Lattice or gelosia multiplication

Early Indian mathematicians of the Indus Valley did multiplication during the Bronze Age – around 3300 through 1300 BC/BCE. They used a method known as the *lattice* (or *gelosia*) method. They performed multiplication using small slate tablets; often using sand or flour, to coat the tablets and then do calculations with their fingers. This impermanence allowed them to use the same tablets over and over. At times they used a white paint like substance to write with.

> *Note*:
>
> The Indus Valley Indians (west side of the Indian continent) were already using our number notation method (symbols 1 through 9) during the Bronze Age. This was long before the fall of the Roman Empire and introduction of the current numbering system to Rome by Fibonacci in the early 1200's AD/CE. Fibonacci also introduced the lattice method of multiplication to the Romans and Western world, in his treatise *Liber Abacii* (*Book of the Abacus*) – published in 1202.

When the method was introduced to Europe it was called the gelosia method. It is named after the iron grill, (now spelled *jalousie*), that was placed over windows in Italy. It is said, Italians installed grills to keep strangers from staring at their wives.

Egyptians and the Rhine Papyrus

By 3000 BC/BCE Egypt become a single nation; formed from two independent nations. They had already developed, and unified their hieroglyphic writing method. During the next 1000 years or so they modified their picture word methods into one known as the *hieratic script*; especially in the area of numeral hieroglyphics. By the last Middle Kingdom and early New Kingdom the Egyptians were performing complex multiplication in the area of geometry.

In 1858 the Scottish Egyptologist A Henry Rhind, purchased a papyrus scroll, about 235 inches long and 13 inches wide, that contained mathematical calculations. It is believed to have been written by the scribe Ahmes and references a papyrus that is more than 200 years older (around 1850 BC/BCE). The papyrus contains eight-seven math problems. Scribe Ahmes is considered the earliest known contributor to mathematics.

Also in the 1850s another papyrus was found – the Moscow papyrus, from the same time period. It contains twenty five problems. Both papyri contain practical arithmetic; like how to build the pyramids.

Since we no longer write in hieratic script, the actual multiplication methodology will not be explained. This section is simply here to reference ancient multiplication.

Indian mathematician Brahmagupta and gomutrika

Brahmagupta wrote two important treatises, or books, in 628 and 665 AD/CE, titled, *Brahmasphutasiddhanta* and *Khandakhadyaka*. Both books addressed the topics of mathematics and astronomy. In the *Brahmasphutasiddhanta* he defined zero and some of its properties when working with operations. He further defined a method of multiplication similar to the way we perform multiplication today. He called it "*gomutrika*"; it comes from the Hindu root word *Gomutra*, which translates as "cow urine." It is believed to literally translate to "trajectory of a cow's urine".

This method utilizes horizontal and vertical numbers for multiplication.

Brother Juan Diez's" Sumario Compendioso" and New World multiplication

Another type of multiplication has its roots in Spain's New World (North America). It is the first mathematical book published in North America.

It was written by Brother Juan Diez and titled, "*Sumario compendioso de las quentas de plata y oro que in los reynos del Piru son necessarias a los mercaderes y todo genero de tratantes Los algunas reglas tocantes al Arithmetica. (1556)*." The title translates to "*Comprehensive Summary of the counting of silver and gold, which, in the kingdoms of Peru, are necessary for merchants and all kinds of traders*." Br. Diez was a Spanish priest who came to Mexico with Cortez in 1519. In this book he describes a unique way to multiply two numbers, where multiplication is performed from left to right. Today it is often referred to as "*Spanish colonial mathematics*"; which references a book by the same name by Ed Sandifer.

Trachtenberg's methods of multiplication

Jakow Trachtenberg developed mental calculation techniques for multiplying, known as the Trachtenberg System. He devised his method while held in a Nazi concentration camp. By focusing on mathematics he was able to maintain his sanity during those horrific times he experienced and witnessed.

Trachtenberg eventually managed to escape from the concentration camp; with the aid of his wife, Countess Alice. She pawned her jewelry to bribe the guards when she learned that her husband was to be executed the next day.

Eventually they went to Switzerland, where he refined his system of mental math. In 1950, Trachtenberg founded the Mathematical Institute in Zürich where he taught his methods. The Institute is still considered a leading school in advancements of unique educational methods in the fields of mathematics and science.

It is based on the simplest of math functions – adding two single digit numbers, halving and doubling a number, and possibly adding a five to the solution.

You already used one of these methods when multiplying by eleven (11). There are other methods for mentally multiplying by one through twelve and beyond.

This book is not intended to cover all the information associated with Trachtenberg's mental methods and will not be covered in this section. Again, it is part of the fascinating history of multiplication.

Curious methods of Multiplication

The history of multiplication shows many different ways to multiply numbers. Some of these methods can still be used today. More importantly, some offer an alternative way that can simplify the process of multiplication.

What follows are several methods used throughout history and may be worth studying.

Lattice, or gelosia Multiplication

As pointed out, Indian mathematics date back to the Indus Valley civilization around 3300 through 1300 BC/BCE. They used a method known as the *lattice* (or *gelosia*) method. Some people call this cell manipulation multiplication.

This method was publicized as early as 1010 AD/CE by the Persian scholar, Karaji (ka-ra-yee) in his book titled, *Kafi fil Hisab*, (Book of Satisfactions).

The figure below shows an example of the lattice method using 132 x 247. Notice that 132 is written across the top, while 247 is written from top to bottom along the right side.

To understand how this method works, we will use the example presented here.

Once you have written the numbers that you wish to multiply by (132 and 247) you will draw diagonal lines through all the squares.

To perform the multiplication, follow these steps:

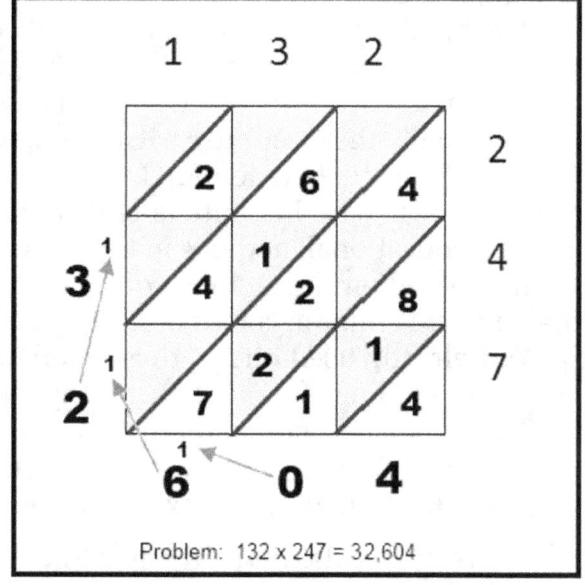

Problem: 132 x 247 = 32,604

1. Starting with the top left square, multiply 1 x 2 and put the answer in the first square. If that answer is a single digit, place it in the white area.
2. Move to the next square and multiply 3 x 2, placing 6 in its square.
3. Repeat step 2 until all the squares are filled with the solution to the individual multiplications.

   Note:
   If the resulting answer is greater than 9, place the tens value in the gray area of each box.

4. Starting at the bottom right, add each diagonal column of numbers up and place their values under the column.

   Note:
   Start at the bottom right most diagonal column, reading up the column you see there is only one cell – 4, so place the 4 under the column. Move to the next column (1, 1, 8); add these values and get 10. Place the 0 under the column and carry the 1 (tens position) to the next column. Add that column (1, 7, 2, 2, 4) you get 16. Place 6 under the column and carry the 1 forward to the next column. Repeat these steps for all remaining columns – placing 2 and 3 under their columns after adding and carrying. The answer is then read from top left to bottom right (down and to the right) = 32,604.

The line crossing method

This is an interesting way to do cross-multiplication. It requires drawing lines on a sheet of paper to represent each number. Then drawing arcs around groupings of numbers and finally counting the intersecting lines in each arc area.

Its origin is unknown; it has been found in books in Asia dating back 200 years. Some believe its origin is from the Mayan civilization; others from India. Wherever it has its roots, it offers an interesting way to perform multiplication graphically.

Here is an example 23 x 41. Using the example, 23 x 41, you draw two lines across the top of the paper (at around a 45 degree angle) to represent the 2 of the 23; then three below them to represent the 3 in 23. Then draw four lines at right angles to the other two lines, crossing them and finally one line to the right of the four lines, again intersecting the lines for the 2 and 3 —one line represents the 4 and the other 1 in 41.

With the lines drawn, separate the top and bottom grouping by drawing an arc around each grouping. Count the intersecting lines in each grouping. Bottom shows three (write a 3). Top shows 8, write 8. Now count the remaining intersecting lines between the two groupings – 12 and 2; which equal 14 – write 14. Move the 1 (10 of 14) to the top grouping, adding it to the 8: 8 + 1 = 9. Now write the answer down– 943!

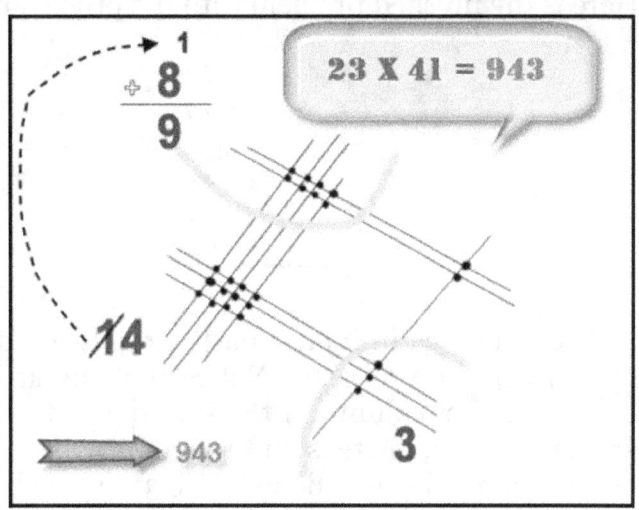

The Gomutrika method

Another method was first used by the Indian mathematician, *Brahmagupta*, in the seventh century. It is known as *gomutrika* ("trajectory of a cow's urine".)

Sorry!

The method he demonstrated is very similar to how we perform multiplication problems today. The difference is how the problem is written down. The top number is written horizontally; however, it is written down the same number of times that there are digits in the second number that is being multiplied. It is also written offset from the previous line (above it). The second number is actually written diagonally with the highest value at the first line and the subsequent numbers below it. Then the actual math is performed.

Here is an example of a multiplication problem using the gomutrika method. The problem to be solved is *317* x *243*:

```
2       3 1 7
4         3 1 7
3           3 1 7
```

Notice that one of the numbers (called a factor) is written horizontally and repeatedly along a skewed line for each digit of the second factor. The second factor (number - 243) is written down the left column of the page – one digit for each horizontal number.

To solve this problem you multiply each single value from the second factor (2, 4, and 3) by 317 and place the resultant answer below the number being multiplied. Here is the finished problem with the process highlighted in italics:

```
2       3 1 7
4         3 1 7        << = = 317 offset by one digit
3           3 1 7      << = = 317 offset by two digits
          6 3 4          << = = 2 * 317 under first 317
          1 2 6 8        << = = 4 * 317 under second 317
    +       9 5 1        << = = 3 * 317 under third 317
          7 7 0 3 1    << = = the ANSWER!
```

The notations in italics (on the right) are added to document the steps, and are not part of the problem. Notice that the answer from (317 x 2), 634, was placed below the first column of the first digit of 317. (If the resultant answer had been greater than 10, the tens unit would have been placed to the left of the top column by one column. This was done in the 317 x 4 column; the 1 was placed to the left of the initial starting column.)

Note:
Although the alignment in this example goes left to right, he also wrote variations with the shift moving to the left from row to row.

So as the problem above demonstrates, the process is very similar to how we multiply today. When the Indian mathematician, Brahmagupta, designed this method in the seventh century he probably didn't anticipate it would still be used 1,400 years later.

Brother Juan Diez's" Sumario Compendioso" method

This method, *Brother Juan Diez's" Sumario Compendioso"*, was created by a priest named Brother Juan Diez. He arrived in Mexico in 1519 with Cortez, and was charged with creating a way to count all the silver and gold that was collected from the kingdoms of Peru and was to be used by the merchants in the new world.

He devised a new way to perform the multiplication calculations.

The method can be a bit difficult to grasp. The best way to explain his method is to demonstrate its usage. Here is the problem 875 x 978.

> *Note*:
> Multiplication starts from the left, using a slash between the numbers being multiplied together. The use of a backward slash sign does not represent division in the problem.
>
> <u>875\978</u>

Looking at this problem, you solve it by first taking the two left most values of each number (8, 9) and multiply them together (actually 800 and 900). Then place your answer, 72 below the 8 of 875:

$$\underline{875\backslash978}$$
$$72$$

Notice that the zeros are left off; similar to the way we leave them off today when we start from the other side.

Next the 800 is multiplied by the 70 in 978. Since this has one less power of ten than the previous multiple, it will start one column to the right, so the 5 of the 56 goes under the 2 of the 72, but since there is nothing in the third column, the 6 goes at the top of this column, next to the 2 of the 72.

$$\underline{875\backslash978}$$
$$726$$
$$5$$

Continuing this process, the next multiplication is between the 800 and the 8 in 978, which is again one less power than the previous multiple, and so the 6 in the 64 must go in the column under the 6 of the previous 56. The 4 will go at the top of the next column since there is nothing in that column yet.

$$\underline{875\backslash978}$$
$$726\ \ 4$$
$$56$$

Understanding the process yet? Continuing, all three digits of 978 have been multiplied by the 8(hundred). Now we move on to the 7 (70) in 875 and multiply it across. Since the 70 times 900 will have only one less place than the 800 times 900, it will start in the second column also, so the 6 of the 63 will go below the 2 and 5 in the second column, and the 3 will go under the pair of sixes in the third column.

$$
\begin{array}{l}
875\backslash978 \\
726\ 4 \\
56 \\
63 \\
\end{array}
$$

Now we move on to 7 (from 875) times 7 (70 from the 978) and the answer 49 starts by placing it in the third column (under 6, 6, 3), and the 9 will go under the 4 in the fourth column.

$$
\begin{array}{l}
875\backslash978 \\
726\ 4 \\
56\ 9 \\
63 \\
4 \\
\end{array}
$$

From here you should be able to continue the process on your own. Here is the completed problem.

$$
\begin{array}{l}
875\backslash978 \\
726\ 460 \\
56\ 95 \\
63\ 54 \\
4\ 3 \\
+\quad 4\ 5 \\
\hline
855\ 750 \\
\end{array}
$$

Once you have placed all the values from the smaller multiplications within the problem, you just add the columns up to arrive at the answer – 855, 750.

That is all there is to using this method. It does seem a bit difficult to follow initially; but once you work with it a while you should realize that it is an efficient method of doing multiplication!

Notice that it is slightly similar to our own method – only doing multiplication from left to right, vs. right to left in our system.

Calculator Tricks – (Part 4)

Day four offers a few more calculator tricks that appear to be magical:

The 6174 loop

1. Select a four-digit number (do not use 1111, 2222, etc.)
2. Arrange the digits in decreasing order
3. Arrange the digits in increasing order
4. Subtract the smaller from the larger number
5. Repeat steps 2, 3, and 4 with the result, and so on, What happens?

ANSWER: solves to a 6174 loop! EXAMPLE: 3241 becomes 4321 – 1234; 8730-0378; 8532 – 2358; 7641 – 1467; becomes 6174!

Another Birthday display

1. Enter the number 7
2. Multiply by the month of your birth
3. Subtract 1
4. Multiply by 13
5. Add the day of your birth
6. Add 3
7. Multiply by 11
8. Subtract the month of your birth
9. Subtract the day of your birth
10. Divide by 10
11. Add 11
12. Divide by 100

ANSWER: displays birthday M.DD or MM.DD

The Mind Reading Number Trick

1. Think of any positive integer (keep it small for doing computations)
2. Square the number (meaning multiply by itself)
3. Add the original number
4. Divide by your original number
5. Add, oh, I don't know, say 17
6. Subtract your original number
7. Divide by 6

Answer: 3! Algebra behind it is x; x^2; $x^2 + x =$
$x(x+1)$; $[x(x+1)]/x = x+1$; $x+1+17 = x + 18$; $x+ 18 – x = 18$; $18/6 = 3$

115

Simple Math is All You Need – (Part 4)

Here are the rules to determine evenly divisible by 2 through 10.

Determine if a number is even divisible by 2 through 10

Division by 2

Any number that ends in 0, 2, 4, 6 or 8 is evenly divisible by 2. Is 4,566 evenly divisible by 2?

Division by 3

Add the number's digits. If the sum is evenly divisible by 3, then so is the number. So, will 3 divide evenly into 2,169,252? Sum of the digits is 27, and 27 is divisible by 3. (Just do check digit math!)

Division by 4

If the number's last 2 digits are 00 or if they form a 2-digit number evenly divisible by 4, then number itself is divisible by 4. How about 56,789,000? Last 2 digits are 00, so it's divisible by 4. Try 786,544. Last 2 digits, 44, are divisible by 4 so, yes. (Remember 4 = 2 x 2; so divide by 2 twice.)

Division by 5

Any number that ends in a 0 or 5 is evenly divisible by 5.

Division by 6

The number has to be even. If it's not, forget it. Otherwise, add up the digits and see if the sum is evenly divisible by 3. If it is, and even, it is divisible by 6. Try 108,273,288. Even and digits sum to 39 which sum to 12 and divides evenly by 3, so the number is evenly divisible by 6. (Check Digits again!)

Division by 7

Multiply the last digit by 2. Subtract this answer from the remaining digits (minus last digit). Is this number evenly divisible by 7? If it is, then your original number is evenly divisible by 7. Try 364. The last digit is 4; multiplied by 2 = 8. Take the remaining digits 36, and subtract 8 = 28. 28 is evenly divisible by 7, so 364 is also! You can continue the process down to a single number comparison!

Division by 8

If the number's last 3 digits are 000 or if they form a 3-digit number evenly divisible by 8, then the number itself is divisible by 8. How about 789,000? Last 3 digits are 000, so it's divisible by 8. Try 786,120. The last 3 digits, 120, divide by 8 will result in 15 with no carry over, so yes.

Division by 9

Sum the number's digits. If it is evenly divisible by 9, it is divisible by 9. As with the tests for 3 and 6, you can keep adding numbers until you're left with only one digit. (Hmm, Check Digit math!)

Division by 10

Any number that ends in 0 is evenly divisible by 10.

116

Multiplying any same number digits by same number of 9's

There is a method in Vedic math that allows you to quickly multiply any number of 9's by a number having the same number of digits. To quickly multiply any number of the same length as the number of nines, follow these steps:

1. Take the number to be multiplied and subtract 1; placing it below the problem
2. Starting with the left most number, take each digit of that number (original minus 1) and subtract them from 9
3. Place them in order next to the number brought down by subtracting 1
4. Once you have performed step 2 for all digits in the original number (minus 1) you are done

For example:

7	47	840
x 9	x 99	x 999
6 < = = (7 − 1 = 6)	46 < = = (47 − 1 = 46)	839 < = = (840 − 1 = 839)
63 < = = (9 − 6 = 3)	465 < = = (9 − 4 = 5)	8391 < = = (9 − 8 = 1)
	4653 < = = (9 − 6 = 3)	83916 < = = (9 − 3 = 6)
		839160 < = = (9 − 9 = 0)

Each of the three examples above take the number to be multiplied by a multiple of 9 and subtract 1; drop the number down below the line and then starting with the left most digit of the original number minus 1, subtracting each one from 9, appending it to the answer.

For example, 9 x 7: Subtract 7 - 1 (6). Place 6 below the problem. Subtract 9 − 6 (3) and place it alongside 6 – the answer: 63. In the second, 99 x 47; subtract 47 − 1 (46); place 46 below the problem; subtract 9 − 4 (5) place 5 next to 46; subtract 9 − 6 (3) place 3 next to the 5 – the answer is 4653. Here are three for you to try:

EXERCISE 19

69	821	84
x 99	x 999	x 99

- Answers are on the next page -

ANSWERS: (Exercise 19)

6831 (69 x 99),   820179 (821 x 999),   8316 (84 x 99)

See how easy that was?

Of course you can still multiply any number of 9's by adding the same number of zeros (multiplying by 10) to the original number and then subtract the original number from the new number.

But if the number you need to multiply by 9's has the same number of digits as the number of 9's being multiplied by; this method is far easier.

Summing a Sequential Series of Numbers

Ever get a problem like what is the sum of 21 + 22 + 23 + 24 + 25 + 26 + 27 + 28 + 29 or perhaps 107 + 108 + 109 + 110 + 111 + 112 + 113 + 114 + 115 + 116? Solving for these types of problems is actually easy.

If you want to add/sum a sequential series of numbers you just need to multiply the exact middle number by the number of items to be added. For instance:

21+22+23+24+25+26+27+28+29 =

Notice that 25 is the exact middle number,
And there are a total of 9 numbers in the series, so –

25 x 9 = 225

Notice the above sequence had an odd number of items, so it had an exact middle number. What about a sequential series with an even number of items? Simply take the average of the middle TWO numbers, like so:

107 + 108 + 109 + 110 + 111 + 112 + 113 + 114 + 115 + 116

(111 + 112)/2 = 223/2 = 111.5 is the average of the middle two numbers in the series.
The number of numbers in the series is 10 so:

111.5 x 10 = 1115

This is a nice demonstration of how to take a commonly used formula (adding series) and rearrange it to use another formula (multiplication) to solve something that is often tedious to compute by hand.

Hopefully, these few tips will help you when doing mathematical calculations.

Endgame for the Day

Today's closing ends with a couple of logic puzzles for fun. Put your thinking caps on and solve these two problems.

The Puzzle: 5 lines of 10 balls

Place 10 balls (circles) along 5 lines. Each line will have exactly 4 balls along it.

Hint: The stars on the American Flag!

Cupids Arrow

The goal of this puzzle is simple: Fill the arrow's circles with numbers that are evenly divisible by 7 or 13. Here are the rules:

1. Use the numbers 1 through 9
2. Each digit may only be used once
3. Place 1 digit in each circle of the magical arrow
4. Once placed, each pair of digits connected by a line must form a 2-digit number (forward or backward) that is evenly divisible by 7 or 13. The number is NOT the sum of the digits, rather the two digits placed together to form a number.

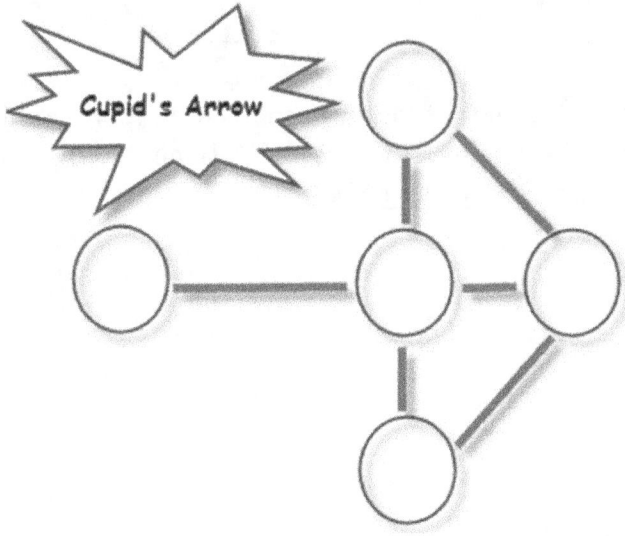

Note

If the numbers in two adjacent circles are 3 and 4 (although they sum to 7) the number would be a 34 or 43; neither 34 or 43 are evenly divisible by 7 or 13 – so it would not be a viable solution.

Here is one possible solution. There are at least four others.

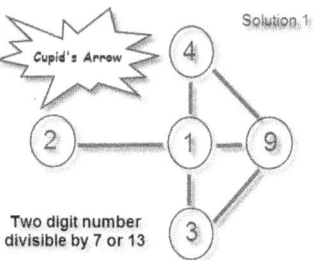

Puzzle developed by Dr. Cliff Pickover.

- Answers are on the next page -

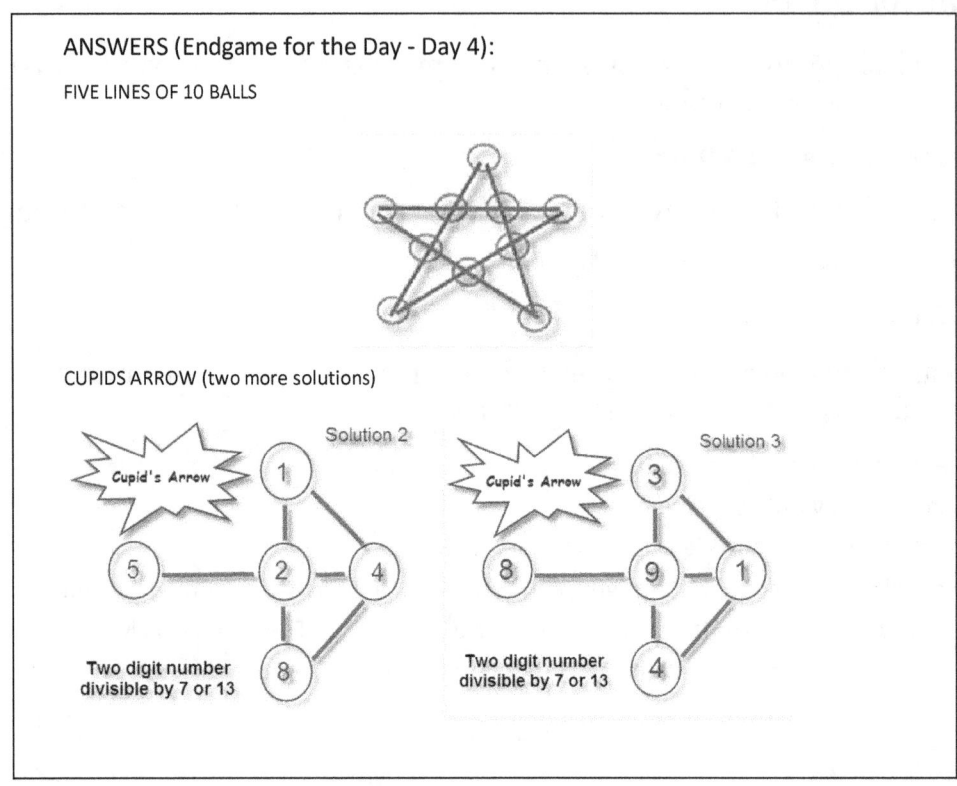

The puzzle, *Cupid's Arrow* was developed by Dr. Clifford Pickover and can be found in his book, *Wonders of Numbers*, published by Oxford University Press, USA (June 2002). This puzzle is used by permission. His official website is <u>pickover.com</u>.

Day Five

WELCOME TO DAY 5 OF MATH 4 2-DAY! The journey continues into the fascinating world of mathematics and the world around us.

Again, if you need to review any material that has been covered over the past four days, you should go back and review it. Today we will discuss a few topics that have made significant impact on the world at large, like Fibonacci numbers and the language of computers. We will also discuss the final part of <u>Avoiding Careless Error Math</u> with multiplication.

Today's Outline

Quotations of the Day
Bit of Fun in Math
    How Much is 2 times 2
    Education = Problems!
Mathematical Curiosities – (Part 4)
    Fibonacci – his Series and the Golden Ration
Avoiding Careless Error Math (Part 3)
Computers and the Math behind them
Calculator Tricks – (Part 5)
Simple Math is All You Need – (Part 5)
    Squaring any number –Mix It Up
    Using 11 – even divisibility and the "Process of 11"
Endgame for the Day
    Mathematical Optical Illusions

Objectives:

(1) Understanding Fibonacci, his contributions to math
(2) Understand how math plays a part in computers
(3) Continue to build upon avoiding careless errors in math
(4) Continue to perform mental math – avoiding the calculator
(5) Work with optical illusions

Quotations of the Day

ಐ · ಐ

Many are ignorant of mathematical truths, not out of any imperfection of faculties, [...], but for want of application ...!

-- *John Locke*

John Locke (1632 –1704) was an English philosopher. He is considered one of the most influential thinkers of his time. His ideas had enormous influence on people like Voltaire and Rousseau, as well as many American revolutionaries. His influence is reflected in the American Declaration of Independence.

ಐ · ಐ

Where there is an open mind, there will always be a frontier.

-- *Charles Kettering*

Charles Kettering (1876 – 1958) was an American inventor and holder of over 300 patents. He was founder of Delco and head of research for General Motors for over 25 years. He was an engineer by profession and founded the Engineers Club of Dayton. His inventions include Freon, the first all-electric ignition system, ethyl gasoline, the pre-cursor to MRIs in medicine, and founder of the Memorial Sloan-Kettering Cancer Center.

ಐ · ಐ

A mathematician, like a painter or a poet, is a maker of patterns. If his patterns are more permanent than theirs, it is because they are made with ideas!

-- *Godfrey H. Hardy*

Godfrey H. Hardy (1877 – 1947) was a popular British mathematician who wrote about number theory and mathematical analysis. He wrote the famous, *A Mathematicians Apology* in 1940 that explained the intra-workings of mathematicians in a common language understood by the lay person. He used the word *apology* in the sense of a formal justification or defense, like Plato, not to plea forgiveness.

ಐ · ಐ

Bit of Fun in Math

Today presents two fun concepts – The first tries to determine just what 2 x 2 really equal and the second demonstrates that education equals problems. Enjoy!

How much is 2 times 2?

Several scientists were asked the following question:

"What is 2 times 2?"

The Philosopher smiles, saying, "*Just what do you mean by 2 * 2?*"

A Physicist looks in his technical references, puts the problem in a computer, and declares, "*it lies between 3.98 and 4.02.*"

The Logician replies, "*Please define 2 * 2 more precisely.*"

The mathematician cogitates for a while and announces: "*I don't know what the answer is, but I can tell you, an answer exists!*"

The sociologist, "*I don't know, but it was nice talking about it.*"

The computer program announces, "*it is exactly '100'.*"

Finally, a Grade school student blurted out, "*4*".

The others look astonished and questioned, "*How did you know*?"

The Student announced, "Simple, I memorized it!"

EDUCATION = Problems!

Mathematical Proof that

> ಜಲ
> EDUCATION = Problems!
> ಜಲ

1. To become educated you need time and money

$$Education = Time \times Money$$

And ...... "Time is Money" ...

$$Time = Money$$

2. Therefore, Education = Money x Money

$$Education = (Money)^2$$

3. Money is "the root of all Problems!"

$$Money = \sqrt{Problems}$$

4. Therefore, Education = $(\sqrt{Problems})^2$

EDUCATION = Problems

Mathematical Curiosities – (Part 4)

Today, we focus on Fibonacci and his famous series of numbers. These numbers, have their roots in a single problem. It was studied by many of his peers and is known as the 'Rabbit Problem.' It has given rise to an interesting series of numbers that have proven invaluable to the world of mathematics.

The Fibonacci Series

This series can be used to demonstrate the beauty and mystery of nature. Many patterns found in nature appear to be driven by mathematics. This series gives us another important mathematical concept – the Golden Section.

Before explaining the series and its impact, a short history of Fibonacci is in order.

A Brief History of Fibonacci and his Contributions to the World of Math

Fibonacci lived during the late 12th and early 13th centuries. His life began in the year 1175, the same year work began on a tower in the city of Pisa, Italy. He was named Leonardo; *no*, not Leonardo da Vinci – this Leonardo was known as Leonardo da Pisa. Like da Vinci, this Leonardo would become famous for his life's work.

Leonardo loved math from an early age. His father, Guglielmo Bonaccio, worked for the government of Italy. For many years, he lived with his father in Northern Africa; known as Barbary (modern day Algeria). Leonardo lived there until his 25th birthday, when the family returned to Italy. While in Barbary, he would spend his days observing people and how they used an unknown method of counting and dealing with mathematics. This system was used in the markets and known as Hindu-Arabic; it used the numerals we use today (0, 1, 2, 3, ...).

He only knew the Roman way of math; using Roman numerals (I, V, X, L, C, D).

Leonardo learned this new system and believed that using ten symbols and position placement was more efficient then the way he had learned as a child.

When he returned to Italy in 1200, he wrote a book about math titled *Liber Abaci(1202)*. In it, he introduced the Latin-speaking world to the decimal number system. He described this new system of writing in this way,

> *"These are the nine figures of the Indians: 9 8 7 6 5 4 3 2 1. With these nine figures, and with this sign 0 which in Arabic is called zephirum, any number can be written, as will be demonstrated. ..."*

Note:
The word 'zephirum' was later shortened to 'zero' in Italian.

Fibonacci and his books

He authored his book under the name "filius Bonacci," Latin for "son of Bonaccio", rather than his given name of Leonardo da Pisa or Lenardo Pisano. Although his books were revolutionary and modernized the mathematics we use today, he refused to write under his real name. Later he would write under another pen name of Leonardo Bigollo (in Tuscany, bigollo means traveler). Today, of course we simply call him Fibonacci (the abbreviated form of "son of Bonaccio".)

When Fibonacci wrote his book *Liber Abaci*, (*Book of Calculation* or *Book of Abacus)*, it changed the world of mathematics. People gave up the Roman numeral method of doing math for the more efficient Hindu-Arabic method. With the invention of the printing press in 1482 it was spread throughout Europe.

Fibonacci is responsible for the modern number system we use today. However, he is more often known for his work that evolved from his problem with rabbits.

Fibonacci's Rabbits

During Leonardo's time, mathematicians often competed with each other by creating mathematical challenges and competitions. In chapter 3, of his book *Liber Abaci*, Fibonacci wrote about a simple series of numbers based on his interest in the reproduction of rabbits. Today this series is known as Fibonacci numbers.

In this puzzle he created an imaginary set of conditions where rabbits could breed, and posed the question, "*How many pairs of rabbits will there be a year from now?*" Or, put in mathematical terms:

Beginning with a single pair of rabbits, if every month each *productive* pair bears a new pair, which will become productive when they are 1 month old, how many rabbits will there be after *n* months?

Fibonacci came up with six sets of conditions –

- Begin with one male rabbit and one female rabbit that have just been born.
- Rabbits reach sexual maturity after one month.
- The gestation period of a rabbit is one month.
- Once reaching sexual maturity, a female rabbit will give birth every month.
- A female rabbit will always give birth to one male and one female rabbit.
- Rabbits never die.

Using these conditions, the following sets of rabbits would develop over time:

Month 0 - At the beginning of the experiment, there is one pair of rabbits.

Month 1 - After one month, the two rabbits have mated but have not given birth. So there are still only one pair of rabbits.

Month 2 - After two months, the first pair of rabbits gives birth to another pair, making two pair in all.

Month 3 - After three months, the original pair gives birth again, and the second pair mate, but do not give birth. This makes three pair.

Month 4 - After four months, the original pair give birth, and the pair born in month #2 give birth. The pair born in month #3 mate, but do not give birth. This makes two new pair, for a total of five pair.

Month 5 - After five months, every pair that was alive two months ago gives birth. This makes three new pair, for a total of eight.

Month 6 - After six months, every pair that was alive two months ago gives birth. This makes five new pair, for a total of thirteen. ... and so on

A graphic demonstrates the problem more clearly,

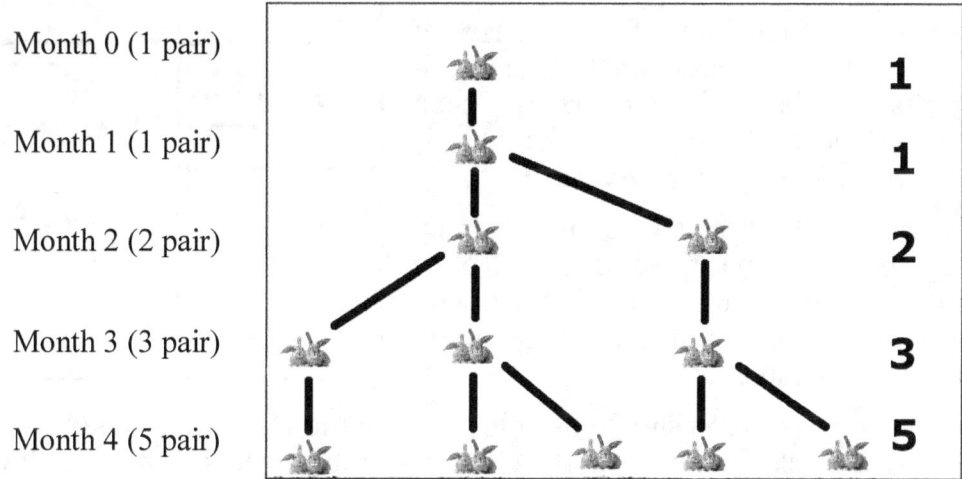

The actual sequence begins: 1, 1, 2, 3, 5, 8, 13, 21, 34, 55, 89, 144, 233, 377, ...

So there you have it! By the end of 12 months (in the 13th month) there will be 377 pairs of rabbits from the original single pair that started in month zero.

This number series became known as the "*Fascinating Fibonaccis*" or simply, the *Fibonacci series* or *Fibonacci numbers*.

Fibonacci Series in Nature

This series of numbers begin with one and one; increasing rapidly by adding the last two consecutive numbers, to form the next in the series. The values of those numbers are seen throughout nature. For example, many flowers have several layers of petals. Some have 3 then 5 petals. 3 and 5 are the fourth and fifth values in the Fibonacci series. This section will focus on Fibonacci numbers in nature.

Fibonacci Rectangles/Squares and Shell Spirals

We can make a rectangular picture of squares based on Fibonacci numbers. To make it, you can draw squares, in sequence, on graph paper. Each square should be placed next to the previous square and will consist of a length equal to a Fibonacci number. To create the rectangles, start with these dimensions – 1, 1, 2, 3, 5, 8, 13, ...

Start with two small (1 x 1) squares, placing them next to each other. Atop of these draw a square of dimension 2 (=1+1) x 2. Draw a new square, touching the 1 square and the 2 square. This new square is 3 x 3 and bordered by another square that is 5 x 5 – alongside the 1 squares and the 3 square. This process continues drawing new squares (counter clockwise) based on the next Fibonacci number in sequence.

This set of rectangles form a series of larger rectangles based on two or more successive Fibonacci numbers (as seen here). Collectively they are known as *Fibonacci Rectangles* or *Squares*.

Here are the first six squares formed from using Fibonacci numbers. Notice that the next square would be drawn above the 3, 2, and 8 thus a 13 x 13 square on the graph paper:

If you continue you should be able to create the 13th and 21st side squares moving in a counter clockwise method. Once this Fibonacci rectangle is drawn, you can draw a spiral starting in the first 1 x 1 box, by placing a quarter of a circle in each square starting with the innermost single sided square. The next page shows how it will look.

<u>Note</u>:
Fibonacci Squares/Rectangles can be seen throughout nature and the art world.

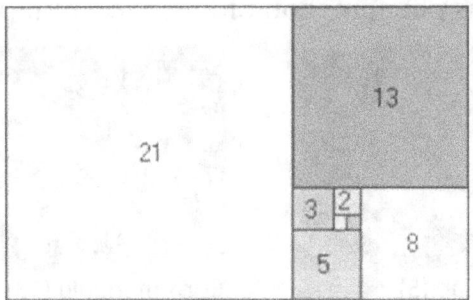
Fibonacci Squares through 21

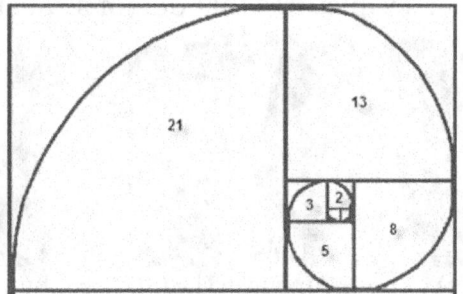
Fibonacci spiral drawn through Squares

The spiral resembles the one above. It's not a true mathematical spiral, since it is made up of fragments that are parts of a circle. But you should get the idea.

This spiral is a good representation of the kind of spiral that appears throughout nature. These types of spirals are seen in the shape of snail shells and sea shells.

Note:
Later you will see how it is also similar to the arrangement of elements in art work.

Below is a picture of a Nautilus shell, by Dylan O'Donnell and released to public domain (2008). It demonstrates application of Fibonacci rectangles/spirals.

Fibonacci Numbers in Flowers and Plants

Did you know that the numbers of petals in a flower are often 3, 5, 8, 13, 21, 34 or 55? Fibonacci numbers! If you look closely at the flowers all around you, many of their petal numbers are Fibonacci numbers – calla-lilies have 1 petal; lilies and iris 3; buttercups 5, corn-marigolds 13 petals; and daisies, often have 34 or 55 petals; finally some sun flowers have as many as 89 petals!

Note:
In addition to the number of petals, each flower often has other parts of the flower that follow the Fibonacci number pattern. For example, many have pistil/stye/stigma and seed pods that also follow the sequence.

Here are a few pictures that demonstrate this phenomenon of Fibonacci numbers:

Calla lily (1)

American Lily (3)

Buttercup (5)

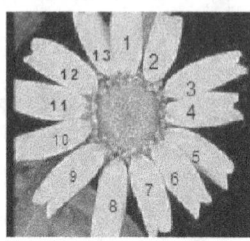
Corn-marigold (13)

In addition to the number of petals, each flower may have other parts that follow some Fibonacci pattern. For example, the calla lily (left) shows 1 petal and 1 stye/stigma. The American lily shows two sets of 3 petals. Finally, the picture on the right demonstrates 13 petals and 1 seed carpel with seeds that spiral in a pattern following Fibonacci numbers as well (more on this later.)

Even seed heads and reproductive parts of plants and flowers demonstrate Fibonacci numbers.

For instance the passion fruit in the picture alongside of this text demonstrates Fibonacci numbers in several ways.

First, there are 2 layers of leaves, comprised of 5 leaves each, 5 anthers, 3 stamens, and finally 1 stigma.

All Fibonacci numbers! WOW!

Fibonacci Spirals

Fibonacci numbers can be demonstrated through the arrangement of seeds on flower heads. The picture to the right is a Coneflower. The actual flower head in this picture is a little more than 2 cm across (slightly less than shown here). The coneflower is a member of the daisy family and has a scientific name of *Echinacea purpura*.

Upon close examination of the seed head you can see it is made up of small "petals" (actually orange in color). They appear to form spirals curving in both directions – to the left and right. If you count those petals spiraling to the right (clockwise) there are 55 spirals. Those moving leftward (counter-clockwise) have 34 spirals. This pair of numbers – 34 and 55 are neighbors in the Fibonacci series.

The spirals themselves are known as Fibonacci spirals.

The same pattern is often seen in many seed and flower heads. Some people theorize it is because this arrangement forms *optimal packing* of the seeds. In other words, no matter how large the seed head, they are uniformly packed at any level; with all seeds being the same size. This eliminates crowding in the center and a sparse arrangement along the outside edges.

These alternating spirals are easily identified by the eye. The numbers of spirals in each direction are (almost always) neighboring Fibonacci numbers!

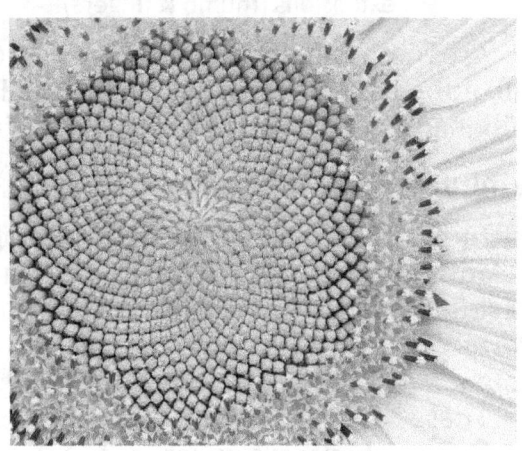

If you look at a sunflower, the seed head is a perfect example. They often have 21 and 34 or 34 and 55 respective spirals. Both are Fibonacci neighbors.

Here is a sunflower – demonstrating the counter clockwise (to the left) and the clockwise (to the right) spirals. If you count them you will find that there are 34 right spirals and 55 left spirals.

As you can see from the picture, there are series of seed spirals that follow Fibonacci numbers, 34 and 55.

Pine cones also demonstrate Fibonacci Spirals. Here is a picture of a pine cone seen from its base (where the stalk connects it to the tree.) There are two sets of spirals. It contains 8 counter clockwise, 13 clockwise spirals

The spirals are obvious when the cone is tightly wound. Once the pine cone opens, it is more difficult to see the left and right spirals. Although they are still there!

If you cross cut a banana or apple open you will see Fibonacci numbers again:

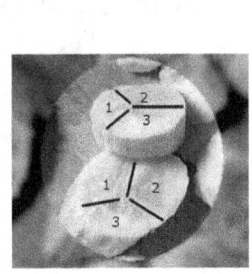
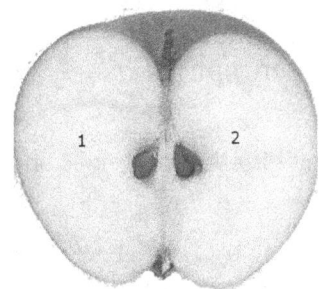
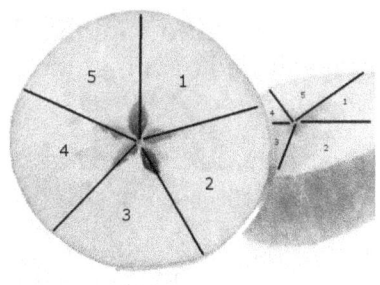

You can see similar Fibonacci numbers in pineapples, other fruits and vegetables.

Fibonacci Fingers?

Have you ever taken a close look at your hands – for instance, you have ...

1 hand that has ...
1 palm and ...
2 knuckles between the ...
3 parts of each finger and there are ...
5 extensions (thumb & fingers)!

Are these simply a coincidence? I don't know the answer. But it is interesting that all of the parts are the first part of the Fibonacci series! Even more interesting is how Fibonacci and his series appear to be linked to the design of the human body.

Fibonacci and the Golden Ratio

In his book, Fibonacci also wrote about two mathematic constants based upon a relationship between two numbers in his series. As the numbers grow within his series he saw a correlation between neighboring numbers. Neighboring numbers formed a constant ratio – these two constants were simple but had a significant impact on math. They impact fields like music, architecture, and others.

These constants can be found in everyday life.

Using Fibonacci's series (1, 1, 2, 3, 5, 8, ...) as a basis, you can obtain the ratio of any two successive numbers by dividing each by the number before it. Doing this you will discover the following series of numbers:

$$\frac{1}{1} = 1; \ \frac{2}{1} = 2; \ \frac{3}{2} = 1.5; \ \frac{5}{3} = 1.66 \ldots; \ \frac{8}{5} = 1.6; \ \frac{13}{8} = 1.625; \ \frac{21}{13} - 1.61538 \ldots; \ \frac{34}{21} = 1.61904\ldots$$

If we continue, the ratio starts to settle down to a particular value, which we call the *golden ratio* or the *golden number*. It has a value of approximately equal to *1.618033988...* This golden number, or golden ratio, is found throughout nature, in architecture and many fields of scientific study.

The Golden Ratio/Golden Section or Phi Φ

The *Golden Ratio*, Golden Section, Divine Proportion, Golden Mean, or Golden Number, is based on a mathematical concept of *Phi*.

Note:

This ratio, 1.618, is known by many different names as shown here. All refer to the mathematical concept of Phi (Φ).

The *golden ratio* is often represented by the Greek letter Phi/phi (uppercase and lowercase Φ or φ). The golden ratio is an irrational constant, approximately equal to *1.618034*.

This *golden ratio*, 1.618034, is represented by the Greek letter *Phi* (Φ). There is also another value closely related to Phi; written as *phi* (φ) with a small "*p*." It is made up of the decimal part of Phi, namely 0.618034.

So *Phi* (1.618) and its identical-decimal relation *phi* (0.618) are constants derived from the Fibonacci series and have lots of interesting properties.

Many great architectural structures in the world are based on this golden ration. These include the Greek Parthenon and the Great Pyramid of Giza.

Famous works of art are based on the golden ratio, including the works of Michelangelo, Raphael, and Leonardo da Vinci; including his famous Mona Lisa. The only difference is that they called it Divine Proportion.

So what is Phi (Φ) and phi (φ)?

Phi (Φ) is a mathematical constant that is approximately equal to 1.618. This concept of Phi in mathematical terms is described as, "Any linear expression (a line, a duration in time) divided by *Phi* will effectively demonstrate 'Golden Section'."

Take a line and divide it so that the ratio of the large piece ("*a*") to the whole line ("*a* + *b*") is the same as the ratio as the small piece ("*b*") to the large piece ("*a*"):

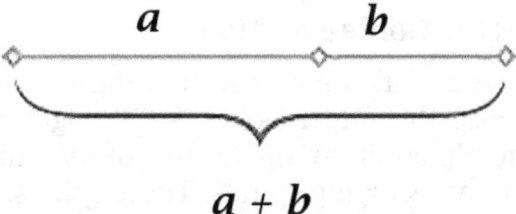

Using the figure above you can conceptualize a correlation between the parts. For example, "*a* + *b*" is to "*a*" as "*a*" is to "*b*". This is only true where "*a* + *b*" is 1.618 times "*a*" and "*a*" is 1.618 times "*b*". Equally, "*a*" is 0.618 times "*a* + *b*" and "*b*" is 0.618 times "*a*". So this means that "*a* + *b*" is 161.8% of "*a*" and "*a*" is 161.8% of "*b*"! Here it is expressed algebraically:

$$\frac{a+b}{a} = \frac{a}{b} = \Phi$$

Conversely it also means that "*a*" is 61.8% of "*a* + *b*" and "*b*" is 61.8% of "*a*"!

> <u>Note</u>:
> The math symbol φ in the formula represents *Phi*, in this case. It is also used to represent its close relative, *phi*.

You can use the numbers 809 for "*a*" and 500 for "*b*"; then plug these values into the above formulas. Working them out, you will see the relationship of both ratios give you *Phi* (1.618).

As pointed out earlier, "*phi*" *(φ)* is a constant related to *Phi*. The mathematical relationship is similar to *Phi*, in that the decimal value is identical to its relative, "*phi*" – 0.618. It is expressed algebraically as:

$$\frac{a}{a+b} = \frac{b}{a} = \varphi$$

So *phi (φ)* is a mathematical constant equal to approximately 0.618

Both *Phi* and *phi* are constants derived from the 21st or greater iteration of Fibonacci's series.

There's something fascinating about these proportions: 1.618 and 0.618

These same proportions, some believe, were used to design you: your hands, face and body! Is this a natural phenomenon? According to Fibonacci, the answer is NO!

The Fibonacci number, *Phi*, is more than a mathematical curiosity; it is one of the most amazing and persistent numbers in the world. Some mathematicians believe that *Phi* and *phi* can be used to support creationism; that it isn't just the result of some random processes.

The Human body and the Golden Section

When someone makes a fist, their fingertips form a spiral (not a circle.) One reason for this is because the finger bones are straight along the back side and curved on the palm side. In addition, the lengths of the finger bones are related in a way often seen in naturally occurring spirals. These spirals, and their separate parts, are related to Fibonacci numbers. In the case of the finger bones the ratios from finger tip through palm are 2, 3, 5, and 8!

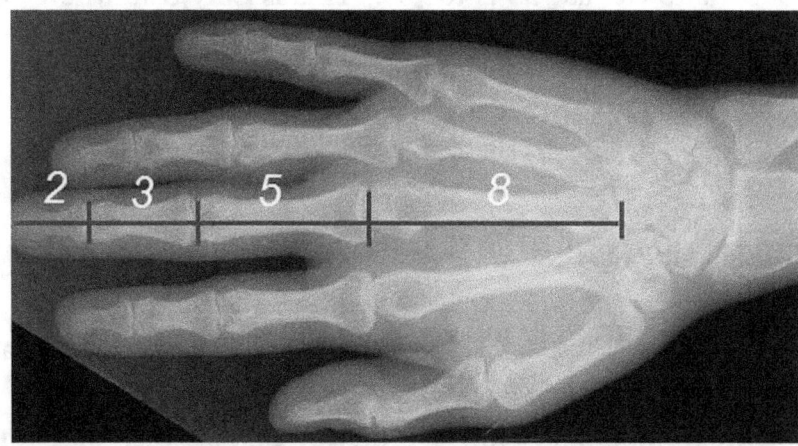

By the way, using this scale, your fingernail is also 1 unit in length. Curious enough you have 2 hands, 3 sections on each finger, each hand having 5 digits, and there are 8 fingers. Again, Fibonacci numbers! Your hand also creates a golden section in relation to your arm, as the ratio of your forearm to your hand is also 1.618, the Golden Section or Divine Proportion.

The proportions of your face are often based on the same ratio. Some scientists suggest that beauty is defined by how closely facial features relate to the Divine Proportion.

The human face has many examples of Divine Proportion. The head forms a golden rectangle with the eyes at its midpoint. The mouth and nose are each placed at golden sections of the distance between the eyes and the bottom of the chin.

Even other parts of the human body appear to be based on Divine Proportion

Divine Proportion and Art

The artists of the Renaissance period knew of the number Phi as the *Divine Proportion*. They used this ratio in their art. There are golden rectangles, golden ratios and even golden triangles.

Father Luca Bartolomeo de Pacioli, or simply *Paciolo*, an Italian mathematician and Franciscan friar, said, "Without mathematics there is no art." Interestingly he was a close friend of Leonardo da Vinci and is credited with being the "*Father of Accounting*"

Looking closely at Leonardo da Vinci's's *Mona Lisa*, you can overlay a Fibonacci rectangle and see how Fibonacci numbers are demonstrated.

This Golden Rectangle is formed from Fibonacci squares –1, 1, 2, 3, 5, 8, 13, and 21.

As the figure demonstrates, the initial rectangle is formed by 1x1 squares aligned vertically along her noise. This starts a perfect Golden Rectangle with the base found alongside the right wrist and moving to her left elbow; finally extending vertically to the top of her head.

The edges of the squares inside this Golden Rectangle actually align themselves along important focal points of the woman: her chin, her eyes, her nose, and the upturned corner of her mysterious smile.

Some believe that Leonardo incorporated mathematics in his art. This painting seems to support that belief!

Golden Triangles and Golden Pentagrams

The number phi φ is seen in geometry, especially in art that contains pentagonal symmetry. Mathematically, the length of a pentagon's diagonal is *phi* times its side.

The golden triangle can be seen an isosceles triangle and the golden pentagram (5 point stars) has ten isosceles triangles: five acute and five obtuse isosceles triangles.

The *Mond Crucifixion* (or the Crocifissione Gavari) by *Raphael*, (1483 – 1520) demonstrates the golden triangle and golden pentagram – mathematically based on Fibonacci.

Raffaello Sanzio da Urbino, simply Rafael, was an Italian painter/ architect of the High Renaissance.

He also based many of his paintings on Fibonacci numbers, both *Phi* and *phi*.

Leonardo and Raphael aren't the only Renaissance Period artists that used Fibonacci numbers and their related geometric designs. Michelangelo was also known to have used Fibonacci proportions in many of his sculptures and painting. These include his famous sculpture, known as *David*, which conforms to the golden ratio and others.

Michelangelo's famous painting the *Holy Family*, demonstrates a golden pentagram. The painting shows Mary receiving her child from Joseph. To the right, in the background, is John the Baptist standing in a pool used for baptism. The golden pentagram (star) appears to have been used for positioning the principal figures – they are aligned with a pentagram.

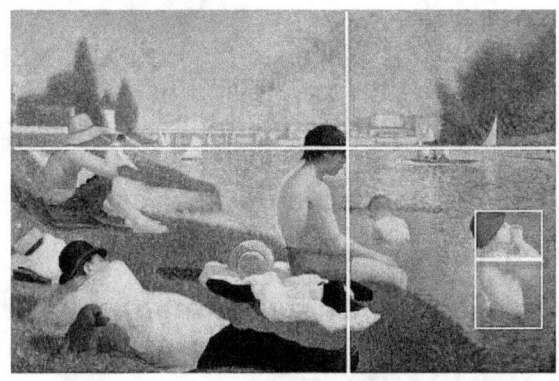

These mathematical relations were not limited to the Renaissance period. Examples are also seen in the works of Rembrandt (1606-1669), Joseph Mallord William Turner (1775-1851), Georges Seurat (1859-1891), and even Salvador Dali (1904-1989).

Here is one of Seurat's paintings titled, *Bathers*. It demonstrates both golden subdivisions and rectangles.

Architecture and Fibonacci

Although Fibonacci published his series in 1202, the use of many of his ideas is seen in much earlier architecture.

Many believed that the ancient Egyptians were the first to use mathematics in their culture. It is obvious that they believed in the magical properties of the golden ratio and used this ratio in the design of the great pyramids. For example, the pyramids of Giza have a ratio of the slant height of the pyramid to the distance from ground center (half the base dimension) of 1.61804 ... (versus 1.61803398 ...) – off by only one unit in the fifth decimal place of Phi!

Even the ancient Greek temple, the Parthenon, is believed to have many of its proportions based on the golden ratio. Its frontage, as well as elements of the façade, appears to be designed based on golden squares/rectangles (non-traditional pattern.) This seems to suggest that the architects were aware of the golden ratio and employed it in their designs.

In closing, Fibonacci numbers can be found everywhere in nature. It is used in architecture and art. Fibonacci should be remembered for more than his series that begins with, 1, 1, 2, 3, 5, 8, 13, 21, 34, 55, 89, ...

He brought Europe and the modern world the number system that we use today – the Arabic numbering system. Finally, his number series gave rise to Phi and phi. In turn these famous ratios are instrumental throughout nature, including human design, the most famous works of art and architecture.

.

Avoiding Careless Error Math (Part 3)

During days one and two you worked with check digits to confirm the answer to your addition and subtraction problems. You probably have a full understanding of check digits and how they work in addition and subtraction.

You have also used them to verify if a number is evenly divisible by 3, 6, or 9.

Verifying Addition and Subtraction problems using Check digits

Having used check digits previously, you should be comfortable with them and understanding their importance. Before using check digits for multiplication, a quick review of check digit use is in order.

Reminder: Creating a check digit

To create a check digit, you sum the individual digits of the number, in non-positional value, with the others in the number. If the resultant answer is a single value between 1 and 9 then you have your check digit for that number. If the number you create by adding all the individual values is more than 9, you continue to add those individual digits until you have a single check digit.

> *Note*:
> You cast out 0's and 9's when summing the individual digits of a number.

For example, 8,739's check digit is 9 (8+7+3+9 => 27 => 2+7 = 9). The number 78,210,914's check digit is 5 (throw out 7+2, 8+1, 0, and 9 and you are left with 1+4).

Check digits with Addition/Subtraction

Before performing the actual addition or subtraction problem, you create a check digit for each number. Then create a comparison check digit by adding /subtracting the individual check digits and bringing that sum down to a single check digit.

> *Note*:
> When subtracting the check digits to create a check digit comparison you may have to take an additional step. If the top check digit is less than the number being subtracted from it (lower number); you need to, first, add 9 to the top digit *before* subtracting the two numbers.

Then you do your addition or subtraction and create a check digit for your answer. Finally, you compare the answer check digit with the comparison check digit. If they match (are the same) then the answer is correct and you move on.

Using Check Digits in Multiplication

Check digits can help even more in multiplication. They are used to verify the correctness of an answer and can also be used to help fix incorrect answers!

If your answer is wrong, check digits can help you determine which line of your multiplication is incorrect and help you fix it. That is some powerful stuff!

Multiplication is just shorthand for addition

It is important to remember that multiplication is simply a short cut method of performing repetitive addition. For instance 25 x 7 is the same as saying:

"Add 25 together (25 to itself, over and over) 7 times"

Using this fact, if a person doesn't know what 25 x 7 is, but does know 25 x5 they can still solve the problem in their head. There are two ways –

- 25 x 7 is the same as (25 x 5) + (25 x 2) or 125 + 50 = 175
- Also the same as (25 x 5) + (25 + 25) or 125 + 50 = 175

Since multiplication is a shorthand method for addition, when using check digits in multiplication you will follow a similar process as in addition and subtraction.

Using check digits in multiplying

The best way to understand check digit math for multiplication is just doing it; use this example: 78 x 35

Create a check digit for each number – multiply the check digits together and create a comparison check digit – solve the problem and create a check digit – compare them.

$$
\begin{array}{ll}
\quad 78 & \quad (6) \\
\underline{x\,35} & \underline{x\,(8)} \\
\quad 390 & \quad 48 ==>> (3) \\
\underline{+\,234} & \\
\quad 2730 \;\; (3) & \quad \text{Match?}
\end{array}
$$

The problem check digits match – so you can assume that the answer is correct.

This is a similar process that you used to perform check digit math in addition and subtraction. Did you notice that once you created your check digits for each number, you multiplied the check digits together to arrive at the check digit comparison number? Since the number was greater than 1 through 9, the answer (48) was solved down to a single check digit.

To use check digits in multiplication, simply follow these steps:

1. Perform check digit math for each number of the problem (78 equals 6/35 equals 8)
2. Do the multiplication of these two check digits (6x8 = 48)
    If the resulting answer is greater than a single check digit (yes – 48), then simplify down until you have a single *comparison check digit* for the multiplied digits (4+8=12 == > 1+2=3)
3. Perform the multiplication problem
4. Create a check digit for the multiplication problem's solution (3)
5. Do they match? (yes?) the answer is correct.

That is all there is to using check digits to perform multiplication.

Using Check Digits to fix an Error in Multiplication

After performing check digit math on your multiplication problem, you may have an error. Don't fret! Check digits can help you figure out where you made a mistake.

When doing a multiplication problem you write a new line beneath it for each number you are multiplying by. Looking at the example below, you create a line for times 5, then underneath it for times 2 (actually 20) and finally times 3 (300).

Once the multiplication is done; the three lines are summed to get the answer. Each part of you solution can be checked for accuracy using check digits.

By performing check digit math on each *interior portion* of the multiplication problem, you can see which line in the problem has a mistake in it.

To perform check digit math on the interior portion of your problem, you take the problem apart and create a check digit for each internal number.

To demonstrate this process, consider the following problem as an example:

$$
\begin{array}{r}
758 \\
\times\ 325 \\
\hline
3780 \\
1516 \\
+\ 2274 \\
\hline
246340
\end{array}
\qquad
\begin{array}{l}
(2) \quad \text{<== check digit} \\
\times\ (1) \quad \text{<== check digit} \\
(2) \quad \text{<== comparison check digit ANSWER} \\
\\
\textit{Note}: \text{Check digits do not match!}
\end{array}
$$

$$(1) \quad \text{<== check digit}$$

Using the above problem, there is an error in the calculations. A mistake is either in the addition or when multiplying 758 x 325. First check to see if we made a mistake in our addition (using check digits). If the addition is OK, we are ready to move to the next step – check the interior portion of our problem using check digits.

Looking internally we can create an interior check digit for each line of our calculations. Once you create a check digit for each line you need to create an internal comparison check digit to use for comparison against each line.

As the example below demonstrates, we create an *internal check digit* for each line of the internal multiplication (check digits 9, 4, and 6). Then we create a new *interior comparison check* digit for each line in the problem. It is created by taking the multiplier digit (units, tens, hundreds) of each line and multiplying that digit by the check digit from the original top line check digit of the initial problem (2).

$$
\begin{array}{rl}
758 & (2) < = = \text{The TOP LINE check digit} \\
\underline{\times\ 325} & \overset{\text{match?}}{\downarrow \qquad \downarrow} \\
3780 & (9);\quad (2) * 5 = (1) \quad \text{INTERIOR number is NOT same, so WRONG} \\
1516 & (4);\quad (2) * 2 = (4) \quad \text{INTERIOR number is same – so correct} \\
\underline{+\ 2274} & (6);\quad (2) * 3 = (6) \quad \text{INTERIOR number is same – so correct} \\
246340 &
\end{array}
$$

Using the problem, you can see each *interior comparison check digit* is created (1, 4, 6). To create the comparison check digit for the first internal line (top, 3780) we multiply 5 (unit value of number 325) times (2) [original top number check digit], getting an *interior comparison check digit* (1). This process continues for the second internal line *interior comparison check digit* –(2) * 2 = 4 as the second line *interior comparison check digit*. Finally the third line's digit is created (2) * 3 = 6.

Comparing the three check digit lines we immediately see that only the first line is wrong [9 ≠ 1] – redo the math for that line!

Re-doing line one, we observe that, the first line should have been 3790 instead of 3780! Here is the solution:

$$
\begin{array}{rl}
758 & (2) \\
\underline{\times\ 325} & \underline{\times\ (1)} \\
3790 & (2) \quad \text{New line value instead of 3780 before} \\
1516 & \\
\underline{+\ 2274} & \\
246350 & \quad \text{Check digits Match!} \\
& \quad \textbf{The answer is CORRECT!} \\
(2) & \quad \text{<== check digit for problem}
\end{array}
$$

So check digits can even be created to figure out where you made a math error in your multiplication problems.

That is powerful!

MOD (7) Residue Double Check Method

Check digit math for verifying your answer in multiplication works the same as for addition and subtraction. It is easy to make a mistake when you do multiplication. It is possible, although unlikely, for those mistakes to give a false positive for the answer. One way to limit errors is to do internal check digit verification.

If you need to re-verify your work, an additional check method can be used to confirm your answers. It is known as the *MOD-7 Residue* method.

> _Note_:
> The term MOD is an abbreviation for modulus.

Understanding the MOD 7 function

This function is extremely useful in math; especially in computer math. It takes any integer (*n*) and divides it by some other integer (*i*) and reports or returns the integer (*whole number*) *remainder* value *only* from the division operation.

Although we are working with *n* MOD 7, MOD can be used with any integer to determine the remainder integer.

BACKGROUND

The mathematical term *MOD*, short for *Modulus*, comes from Modular Mathematics. Modular math is the study of mathematics fathered in 1801, by Johann Carl Friedrich Gauss, a German mathematician.

Place in number theory terms, modulus means (typically unclear math terms), '*The expression of all numbers in terms of a known multiple and a residue (remainder), where the residue is equal to or less then the known multiple.*'

Some examples of *n MOD i* function:

1 MOD 4 = 1 (1/4 = 0 remainder 1)   37 MOD 12 = 1 (37/12 = 3 remainder 1)
3 MOD 4 = 3 (1/4 = 0 remainder 3)   23 MOD 3 = 2 (23/3 = 7 remainder 2)
4 MOD 4 = 0 (1/4 = 1 remainder 0)   128 MOD 7 = 2 (128/7 = 18 remainder 2)
5 MOD 4 = 1 (1/4 = 1 remainder 1)   42 MOD 7 = 0 (42/7 = 6 remainder 0)
                                                        And so on...

The examples demonstrate that the number (*n*), in front of MOD is *divided* by the number after MOD (*i*). The only part of the answer that is kept is the remainder (it is solved down to an integer - no decimal points) So *128 MOD 7* means to divide *128 by 7*; drop all parts of the answer *EXCEPT* the remainder (2).

Looking at the examples, we see that the MOD function is *ONLY* concerned with the remainder of the division problem.

When using Modulus (MOD), *BOTH* operands must only be integers and the remainder is reported as an integer – not a fraction or decimal!

Understanding the MOD 7 residue method

The MOD 7 residue method (*n MOD i*) only supplies the left-over (remainder) value. It is possible for someone to solve a multiplication problem and due to human error, accidently write two or more digits incorrectly somewhere in the answer. It is also possible that the check digit math problem will show the answer is correct; even though it is wrong!

This is where the MOD-7 residue check method comes in handy. MOD 7 residue performs a MOD 7 calculation on each digit of the multiplication problem AND the answer; then does a comparison to verify that they match.

Note: Create a MOD 7 value for each number of the problem – then multiply those numbers together – solve the problem and create a MOD 7 value for the answer – compare the numbers.

Again the best way to see how it works is to use an example:

$$
\begin{array}{ll}
3,527 & (8) \\
\underline{\times\ 1,778} & \underline{\times\ (5)} \\
6,181,006 & (4) \\
\end{array}
$$

(4)   CHECK DIGITS MATCH

Using this example, the check digit math comparison is good. Therefore the problem appears to be correct. However, I can assure you that the solution is incorrect. It should be 6,271,006 (try it out on a calculator).

Using MOD 7 residue method to verify our work, we see that the MOD 7 values do not match. Here is how it would look:

$$
\begin{array}{lll}
3,527 & [6] & 3527\ MOD\ 7 = 6 \\
\underline{\times\ 1,778} & \underline{\times\ [0]} & 1778\ MOD\ 7 = 0 \\
6,181,006 & [0] & \\
\end{array}
$$

6181006 MOD-7 IS [6]   MOD-7 VALUES do not MATCH

Even though the check digits matched earlier, the MOD-7 residue (remainders) does not. So the answer is wrong.

Of course if you had solved the problem by writing each line down, you could have performed the check digit math on the internal lines and know which lines are wrong. Double check your work and perform a MOD-7 check against the numbers.

How to perform a MOD 7 Calculation

Believe it or not, there is an easy way to perform a MOD 7 calculation against any number. Use the following number as an example: **3459 MOD 7**

To perform a MOD 7 calculation, you start by using the leftmost two digits of the number and perform a MOD 7 operation on them, and move through the number from left to right.

Follow these steps:

1. Divide the left most two digits by 7 ==> 34/7 = 28, residue 6
2. If there is a residue, combine it with the next digit ==> 65
3. Repeat steps 1 and 2 until you have no more numbers to divide by seven and keep the residue (remainder)

Note:

Here is a table of multiples of 7 (multiply by) giving no remainder to use for the closest number:

7 (1), 14 (2), 21 (3), 28 (4), 35 (5), 42 (6), 49 (7), 56 (8), 63 (9), 70 (10), 77 (11), 84 (12), 91 (13), 98 (14)

Following the above steps and using the table as a guide, you can now determine the residue/remainder to the problem 3459 MOD 7:

3,459 Start with left 2 digits   = = >> 34/7 = 4 remainder 6

  659 Continue with remainder and next digits = = >> 65/7 = 9 remainder 2

   29 Continue with remainder and next digits = = >> 29/7 = 4 remainder 1

The final residue/remainder is = = >> **1** (THE ANSWER)

So the answer to 3459 MOD 7 is 1!

The MOD-7 of any number, starts with the left most two digits, divide by 7 THEN add the residue from that calculation to the next digit until you reach the end.

We can now solve for the MOD 7 of each number:

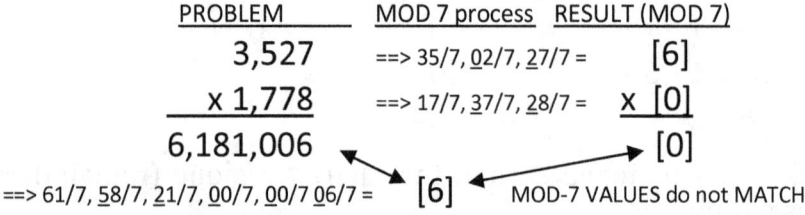

PROBLEM	MOD 7 process	RESULT (MOD 7)
3,527	==> 35/7, 02/7, 27/7 =	[6]
x 1,778	==> 17/7, 37/7, 28/7 =	x [0]
6,181,006		[0]

==> 61/7, 58/7, 21/7, 00/7, 00/7 06/7 = [6]   MOD-7 VALUES do not MATCH

So redo the problem and check again:

PROBLEM	MOD-7 process	RESULT (MOD-7)
3,527	==> 35/7, 02/7, 27/7 =	[6]
x 1,778	==> 17/7, 37/7, 28/7 =	x [0]
6,271,006		[0]

==> 62/7, 67/7, 41/7, 60/7, 40/7 56/7 = [0]   MOD 7 VALUES DO MATCH

That is all there is to check digit math with multiplication!

Computers and the Math behind Them

Modern day computers are based solely on mathematics. They use a system of math known as Base-2. They work on the principle of electrical current either ON (passing through a circuit) or OFF.

To understand the math of computers it is important to review the history of numeral systems and numbers.

A Short History of Numbers and Numeral Systems

A *numeral system* (or officially known as a *system of numeration*) is a method used to represent mathematical notation. It is based on a set of symbols that are used in a pattern that is recognizable by the person performing calculation. There are many different systems of numeration. They are often referred to as a Base-*number*; where *number* represents the number of symbols used to perform mathematical operations. For example, our current system of mathematics is known as Base-10.

Base-1 or Unary System

The simplest numeral system is known as the unary numeral system (Base-1). This system uses a single symbol to represent each digit by writing a corresponding number of symbols for each number. If the symbol * is chosen, for example, then the number seven would be represented by * * * * * * *. This system is often known as the tally system. Here is a tally system found on the *Lebombo* bone dating back 37,000 years ago:

Another example of the unary (Base-1) system is the modern day tally marks - groups of five lines. Here is an example of Base-1 demonstrating values 1 through 5:

There are many Base- systems in use in the world today, including Base-1, Base-2, Base-5, Base-6, Base-10, Base-12, Base-16, Base-20, and Base-60.

Today's Base-10 System

Our everyday math has been in use universally since the early 1400's; introduced in the 1200's. It is formally known as the Hindu-Arabic numeric system, which is based on a combination of 10 symbols or digits – 0, 1, 2, 3, 4, 5, 6, 7, 8, and 9. These ten digits were introduced to Europe by Fibonacci. He learned the system of numeric values while living in Northern Africa in the late 1100's.

Importance of Zero

Zero (0) is a number, with a value of nothing – null- or nil, and a positional digit used to represent a potential number. It plays a central role in mathematics as the additive identity of the integers, real numbers, and many other algebraic structures.

The Mayan concept of zero

The Mayans introduced the concept of zero (0) in the late 4th Century BC/BCE. Their earliest recorded use can be found in the *Mesoamerican Long Count calendar*, a non-repeating, vigesimal (Base-20) calendar used by the Mesoamerican cultures, most notably the Maya. It is known as the *Mayan Long Count* calendar. It is a non-repeating 365 day calendar that begins based on a day 0, August 11, 3114 BC/BCE, and ends in the year 2012. The Maya also performed basic math operations using the concept of 0. In their system, they used a series of dots and bars to represent numbers. The zero was represented by a shell symbol : .

Some credit the Mayans with being the first people to visualize the concept of 0 and its' use as a place holder. However, their influence was limited to Mesoamerica only (South and Central American region).

Zero's FIRST documented use in math computations

It is difficult to follow the history of zero and who first used it in mathematics. Since zero was used as a place holder and a value of nothing – or lack of a value, its usage and the idea of zero didn't instantly appear in the world of mathematics.

For example, by the middle of the 2nd millennium BC/BCE the Babylonians used a sexagesimal (base 60) numeral system. They used a space between the numerals to signify a positional value (or zero). Around 300 BC/BCE they used two slanted wedges to represent a positional 0. A tablet (around 700 BC/BCE) was found in modern day Iraq (about 50 miles south of Baghdad.) It is credited to the scribe Bêl-bân-aplu, who wrote zeros using three hooks, rather than two slanted wedges.

The Babylonian placeholder was not a true zero because it was not used alone.

Zero being used as a number, rather than a symbol for separation, or place holder positioning, is credited to India. The earliest documentation of its' use in calculations was in the 9th century AD/CE. It was used like any other number (1-9) – even in the case of division. The history of zero and its use actually dates back two centuries earlier when the Indian mathematician Brahmagupta defined rules involving zero and negative numbers in the seventh century. He explained that if you take a number and subtract it from itself you obtain zero. His rules for addition which involve zero were:

The sum of zero and a negative number is negative, the sum of a positive number and zero is positive; the sum of zero and zero is zero.

His work and the work of other Indian mathematicians were translated by Islamic and Arabic mathematicians; thus giving rise to the use of zero through the west.

The Islamic mathematician Al'Khwarizmi wrote about Hindu-Arabic numerals (1-9, and 0), around 813 – 833 AD/CE, and is credited with documenting zero as a place holder in positional notation. He is credited with creating the word *algebra*; writing a book on linear algebra and quadratic equations.

> *Note*:
>> Some say that the modern day name "zero" ultimately comes from the Arabic word *sifr*. However, the initial Hindu word 'zephirum' may be the source that was later shortened to 'zero' in Italian.

Hindu-Arabic System

The modern day Hindu-Arabic numeral system was developed in ancient India (historically in Hindus – an area in modern day India, Afghanistan, and part of China/Tibet). This system represents the concept of a pure place-value system and a representation of the concept of 0 (zero). It dates back to the 4th Century BC/BCE. Initially 0 (zero) was represented by a space between numbers.

Despite its Indian origin in the West it was known as *Arabic numerals* only; because of its introduction to Europe through Arabic texts such as Al-Khwarizmi's *On the Calculation with Hindu Numerals* (ca. 825 AD/CE), and Al-Kindi's four volume work *On the Use of the Indian Numerals* (ca. 830 AD/CE). Arabs were the pincipal disseminators of the system.

So there you have it, our Base-10 decimal system is descendent from the Hindu Numeral system, dating back to the early 4th Century BC/BCE. It was in wide spread use around the 7th Century AD/CE by the Arabs. It was introduced to the European culture around 1202 by Fibonacci through his *Book of Calculation, Liber Abaci,* and came into wide spread use in Europe upon the invention of printing in 1482.

Base-2 to the Rescue!

The Base-2 number system, or binary numeral system, represents numeric values using two symbols, usually 0 and 1.

Official mathematical nomenclature defines base-2 as, "a positional notation with a radix of 2." Base 2 is the official system used to represent the planning and implementation of electronic circuitry. This is why the binary system is used internally by all modern digital computers.

A computer is based on the principal of 'reading' the state of two conditions – electricity passing through a circuit (wire) or not. If the electricity is *on* it is represented by a 1 and if *off*, by a 0. Using this concept of ONs and OFFs several circuits can be read and strung together to represent a series of on and off states.

Base-2 is not a new concept; it's been documented since the 8th century; although some suggest it has been around as early as the 3rd century. Roger Bacon described a system where letters of the alphabet can be reduced to a sequence of binary digits in 1605. By the 17th century, Gottfried Leibniz (inventor of Calculus simultaneously with Isaac Newton) fully documented, in his article *Explication de l'Arithmétique Binaire*, a system that used 0 and 1, like modern binary numeral systems.

Binary, Base-2 and Computers

Base-2 is the numbers of computers. Computers use terminology like bit and byte, which are based on binary notation. Computers use binary internally (0 / 1) to represent a sequence (or series) of bits (binary digits). Each series (bits) are a part of a sequence and can represent an operation, instruction, or even a character (A, B, 1, 6) visible on the screen. Each individual *sequence* of binary states (0s and 1s), called a byte, is mutually exclusive of all other states.

> *Note*:
> A *bit* is a single digit, which has a value of 0 or 1. For example, the binary number 10010111 is made up of 8 individual bits. In turn, those 8 bits in a single series represent one *modern byte*. These bits, when combined, are the basic unit used to store unique information on a disk, in memory, or other storage devices. They are the same units of 'information' that are used when sending information from one computer to another in digital computing.

Computers use bytes to do operations and store information. They can use any numbers of bits strung together to make a byte. The simplest byte is made up of eight bits (1s and 0s). Here is an eight bit byte: (spaces are present for clarity only)

$$1\ 0\ 0\ 1\ 1\ 0\ 1\ 1$$

This byte is comprised of eight individual bits or digits. The computer reads these 'bytes' of data and perform operations based on them. Today's computers can use more than eight bits at a time. For instance 16-bit computers use two bytes at a time, while a 32-bit or 64-bit computer read and use 4 and 8 bytes at a time.

Working with Base-2 in a Base-10 World

Since we perform math in a base-10 world, an understanding of how base-2 works as compared to base-10 (Converting from one to another), is in order.

The base-10 number system uses 10 digits: 0, 1, 2, 3, 4, 5, 6, 7, 8, 9. Numbers above 9, such as 10 or 23, are actually a combination of numbers in the sequence above. In the case of 10 it is a combination of 1 and 0 and 23 is 2 and 3. This forces a positioning of the digits (like 2 and 3) into a given location within a number.

Every time a 9 is reached in any column, the value of the column to the left is checked and if it isn't a 9 already, the value is increased by 1 and the current 9 value is converted to a 0. For example the next value after 109 is 110 (the 2nd place - 10's - is increased by 1 and the unit's becomes a 0), 189 becomes 190. This process continues until all values to the right are 9's and then the left most value is converted to a 1 and a new zero is added to the next column to the right – 999 (three) becomes 1000 (four digits).

Another way of thinking about base-10 is when reading from left to right, think of each number being based on exponents (powers). The right most power is 0, the one to its left is 1, next to it is 2, and so on. So the number 234 would be:

$$2 \times 10^2 + 3 \times 10^1 + 4 \times 10^0 = 200 + 30 + 4.$$

In the case of base-10 there are 9 value digits. Looking at the above formula the left most column represents hundreds (10^2), the next tens (10^1) and the last represents units (10^0) This concept is important.

In contrast to using ten digits (0 – 9), base-2 only uses 2 digits – 0 and 1. This means that the three digit number, 111_{10} (base-10) is not the same as 111_2 (base-2). 111, in base-2 represent's the number 7 in base-10! In base-10 we read the number 111 as 'one hundred eleven'. In base-2 we read one, one, one! Quite a difference!

Notice the sub-script 10 or 2 in the numbers 111 above. A subscript 2 (X_2) represents base-2 while a subscript 10 (X_{10}) represents base-10.

Another way of thinking about base-2 is when reading from left to right, think of each number being based on exponents (powers), like in base 10. The difference is the base number will be a 2 instead of a 10. Again, the right most power is 0, the one to its left is 1, next to it is 2, and so on. So the binary number 111_2 would be:

$$1 \times 2^2 + 1 \times 2^1 + 1 \times 2^0 = 4 + 2 + 1.$$

Binary, like base-10 is read from left to right and the leftmost value (of 111_2) is higher than the values of the other two 1s to its right.

Converting Base-2 to Base-10

This table illustrates an 8-bit binary number and the maximum base-10 value that each one in base-2, by position, represents per digit:

Base-2	1	1	1	1	1	1	1	1
Base-10	128	64	32	16	8	4	2	1

Looking at the table above, the top row shows all eight digits (a byte) with values of one – 11111111_2 (base-2). This represents the number 255_{10} (base 10). You get 255 by summing all values under each single bit (1): 128 + 64 + 32 + 16 + 8 + 4 + 2 + 1 = 255 in base-10.

The rightmost digit 1 in binary represents the number 1 in base-10. The next (from right to left) bit 1 in base-2 represents a value of 2 in base-10. If the rightmost first digit is a zero and the left of first digit is a 1 (or 2 + 0, in base-10) the actual number 10 in base-10 represents the value 2 in Base-2.

Base-2	1	1	1	1	1	1	1	1
Base-10	128	64	32	16	8	4	2	1
Number in base-2	0	0	0	0	0	0	1	0
= base-10		2						
Number in base-2	0	0	0	0	0	1	0	1
= base-10		5 (4 + 1)						
Number in base-2	0	0	0	0	1	0	1	1
= base-10		11 (8 + 0 + 2 + 1)						
Number in base-2	0	0	0	1	0	1	1	0
= base-10		22 (16 + 0 + 4 + 2 + 0)						
Number in base-2	0	0	0	1	0	0	0	1
= base-10		17 (16 + 1)						
Number in base-2	1	0	1	0	1	0	1	1
= base-10		171 (128 +0 +32 +0 + 8 + 0 + 2 + 1)						

Using the table above, there are six examples of Base-2 numbers (0 and 1) and their corresponding base-10 values. For example the first example in the above table in base-2 is: 0 0 0 0 0 0 1 0, which is the number 2 in base-10.

Note:
If any specific Base-2 value is set to 0, the Base-10 value is also 0.

To convert from Base-2 to Base-10, you read all bits (from left to right) of the Base-2 number, converting each bit to its corresponding Base-10 value. Then sum all Base-10 values together to convert the number from Base-2 to Base-10.

For instance, using the third example– *0 0 0 0 1 0 1 1*, you would read it from left to right (starting with 0). The first digit that has a value of 1 is under the fifth column from the left. It represents 8_{10}. Moving to the next 1 in the seventh column from the left is 2_{10} and the final 1 column eighth from the left represents 1_{10}. Summing all Base-10 values (8+2+1), for the Base-2 number *00001011* the equivalent value is 11_{10} in base-10   So, to convert from Base-2 to Base-10 sum all corresponding Base-10 values for the 1's in Base-2. So 1001_2 is 9_{10} (8 + 1).

Like numbers in base-10 you don't need to put leading 0s in front of any number. Any number that includes a 0 as one of its internal digits must be shown. For example 00100_2 would be written as 100_2 (4_{10}) dropping the leading 0's.

Computers and Standard ASCII Character Set

So now you can convert a binary number (base-2) into a base-10 number. Computers don't convert binary code to base-10. Rather, it uses a standard scheme that interprets binary. This standard is known as *ASCII,* or *American Standard Code for Information Interchange.* ASCII is a character encoding method based on English characters. ASCII codes represent text. It converts 8-bit, or byte, binary code into plain English.

> *Note*:
> ASCII isn't the only standard used to code binary data. IBM uses its own method for use on its Mainframe and mid-range computers. It is known as *Extended Binary Coded Decimal Interchange Code (EBCDIC).* It is descended from the code used on punch cards.

The 127 values based on the first 8-bits, are converted to English characters which are used by the computer. The two tables that follow show the values from binary 32_{10} ($010\ 0000_2$) through 126_{10} ($111\ 1110_2$). The eighth (leftmost) bit is always a 0.

Binary	Base-10	Character	Binary	Base-10	Character	Binary	Base-10	Character
010 0000	32	SP	011 0000	48	0	100 0000	64	@
010 0001	33	!	011 0001	49	1	100 0001	65	A
010 0010	34	"	011 0010	50	2	100 0010	66	B
010 0011	35	#	011 0011	51	3	100 0011	67	C
010 0100	36	$	011 0100	52	4	100 0100	68	D
010 0101	37	%	011 0101	53	5	100 0101	69	E
010 0110	38	&	011 0110	54	6	100 0110	70	F
010 0111	39	'	011 0111	55	7	100 0111	71	G
010 1000	40	(	011 1000	56	8	100 1000	72	H
010 1001	41	)	011 1001	57	9	100 1001	73	I
010 1010	42	*	011 1010	58	:	100 1010	74	J
010 1011	43	+	011 1011	59	;	100 1011	75	K
010 1100	44	,	011 1100	60	<	100 1100	76	L
010 1101	45	-	011 1101	61	=	100 1101	77	M
010 1110	46	.	011 1110	62	>	100 1110	78	N
010 1111	47	/	011 1111	63	?	100 1111	79	O

Using this table, every time the computer encounters a byte with the value of 65_{10} (or $100\ 0001_2$) it puts a capital "*A*" on the paper or the screen. If it encounters the value 64_{10} (or $100\ 0000_2$) it places the "@" symbol, and so on.

So the computer reads binary values that can be converted to base-10 values. Each value, in turn represents some character, letter, number, or other symbol that is placed on the computer screen or stored on the disk.

The table below continues the representations of 80_{10} through 127_{10}.

Binary	Base-10	Character	Binary	Base-10	Character	Binary	Base-10	Character
101 0000	80	P	110 0000	96	`	111 0000	112	p
101 0001	81	Q	110 0001	97	a	111 0001	113	q
101 0010	82	R	110 0010	98	b	111 0010	114	r
101 0011	83	S	110 0011	99	c	111 0011	115	s
101 0100	84	T	110 0100	100	d	111 0100	116	t
101 0101	85	U	110 0101	101	e	111 0101	117	u
101 0110	86	V	110 0110	102	f	111 0110	118	v
101 0111	87	W	110 0111	103	g	111 0111	119	w
101 1000	88	X	110 1000	104	h	111 1000	120	x
101 1001	89	Y	110 1001	105	i	111 1001	121	y
101 1010	90	Z	110 1010	106	j	111 1010	122	z
101 1011	91	[	110 1011	107	k	111 1011	123	{
101 1100	92	\	110 1100	108	l	111 1100	124		
101 1101	93	]	110 1101	109	m	111 1101	125	}
101 1110	94	^	110 1110	110	n	111 1110	126	~
101 1111	95	_	110 1111	111	o	111 1111	127	DEL

So the computer converts the binary instruction or code to an English character and prints, displays, or uses it as an instruction (like DEL), performing the appropriate action.

Note:
ASCII values 128 through 255 are known as EXTENDED ASCII Code and include symbols like Ç (128_{10} or $1000\ 0000_2$), Æ (146_{10}), ¿ (168_{10}), ▓ (178_{10}), π (227_{10}) and so on ...

Of course a computer does far more than convert binary 8-bit code to ASCII characters. It uses other binary code to perform specific actions.

Now you should be capable of converting binary, base-2 numbers into base-10 numbers.

Hope this helps explain the math of computers. It's all in the 1s and 0s!

Calculator Tricks – (Part 5)

A few more calculations that appear to be magical ...

The Golden Prediction

START with a sheet of paper with 25 lines on it (next page). Follow these steps
1. Write any TWO (2) whole numbers on lines 1 and 2 (First two lines)
2. Add the two numbers together and write the SUM on the next line (line 3)
3. Repeat the process of adding the last two lines together / write the SUM on the next line
4. Repeat step 3 until you have 25 or more numbers on the list
5. Select any number from lines 22 – 25 and divide that number by the previous line number (line 21 – 24). If you select line 24's number, divide it by line 23's number

Answer will be Phi (1.618033989 ...)
INTERESTING FACT: If you divide any of the last five numbers by the next highest number you will get the resulting number 0.618033989 ... (phi)

The 4-2-1 Loop

PRESS Enter after every operation

1. Enter a whole number into your calculator
2. If the number is EVEN, DIVIDE by 2 - - -
3. If the number is ODD, multiply by 3, ENTER, ADD 1
4. Repeat the process in STEP 2 over and over
5. UNTIL you see a pattern ... which is ????

Answer will be a perpetual loop: 4 2 1 – 4 2 1 – 4 2 1

Mystery of 24 and Prime Numbers

Again, PRESS Enter after every operation

A list of prime numbers from 1 through 2000 can be found two pages forward

1. Select any prime number greater than 3
2. Square it – times itself
3. Subtract 1
4. Divide by 24
5. Was it evenly divisible by 24?

Answer: Every prime number will be divisible by 24
Primes are $(6n+1)$ or $(6n-1)$; therefore $(6n+1)^2 - 1 = 36n^2 + 12n + 1$
$-1 = 12n(3n + 1)$ so any prime will be divisible by 24

LINE Put 2 numbers in lines 1 & 2 (think 1 or 2 digits)

LINE	
1	
2	
3	
4	
5	
6	
7	
8	
9	
10	
11	
12	
13	
14	
15	
16	
17	
18	
19	
20	
21	
22	
23	
24	
25	

ADD line 1&2 put on

ADD line 2&3 put on

ADD line 3 & 4 together

Continue process for all 25 lines – add current line value to previous line value & put on the next line. Until you fill all 25 lines!

Select a line number between #22 and 25. Enter in Line # blank. Place value alongside the line number on top of fraction. Place value from previous line as denominator. SOLVE!

Line #___ ——————— = _____

Answer will be (Phi) 1.618033989 ...

PRIMES from 1 to 2000

2	3	5	7	11	13	17	19	23	29	31	37	41
43	47	53	59	61	67	71	73	79	83	89	97	101
103	107	109	113	127	131	137	139	149	151	157	163	167
173	179	181	191	193	197	199	211	223	227	229	233	239
241	251	257	263	269	271	277	281	283	293	307	311	313
317	331	337	347	349	353	359	367	373	379	383	389	397
401	409	419	421	431	433	439	443	449	457	461	463	467
479	487	491	499	503	509	521	523	541	547	557	563	569
571	577	587	593	599	601	607	613	617	619	631	641	643
647	653	659	661	673	677	683	691	701	709	719	727	733
739	743	751	757	761	769	773	787	797	809	811	821	823
827	829	839	853	857	859	863	877	881	883	887	907	911
919	929	937	941	947	953	967	971	977	983	991	997	
1009	1013	1019	1021	1031	1033	1039	1049	1051	1061	1063	1069	1087
1091	1093	1097	1103	1109	1117	1123	1129	1151	1153	1163	1171	1181
1187	1193	1201	1213	1217	1223	1229	1231	1237	1249	1259	1277	1279
1283	1289	1291	1297	1301	1303	1307	1319	1321	1327	1361	1367	1373
1381	1399	1409	1423	1427	1429	1433	1439	1447	1451	1453	1459	1471
1481	1483	1487	1489	1493	1499	1511	1523	1531	1543	1549	1553	1559
1567	1571	1579	1583	1597	1601	1607	1609	1613	1619	1621	1627	1637
1657	1663	1667	1669	1693	1697	1699	1709	1721	1723	1733	1741	1747
1753	1759	1777	1783	1787	1789	1801	1811	1823	1831	1847	1861	1867
1871	1873	1877	1879	1889	1901	1907	1913	1931	1933	1949	1951	1973
1979	1987	1993	1997	1999								

Simple Math is All You Need – (Part 5)

Before working with a few new ideas, a couple of old ones need reviewing:

Multiply any number by 10

WHOLE Numbers

x 10 – add ONE zero   x 10% - move decimal ONE to LEFT
x 100 – add TWO zeros   x 1% - move decimal TWO to LEFT

45 x 10 = 450   45 x 10% = 4.5
45 x 100 = 4500   45 x 01% = 0.45

DECIMALS Numbers

x 10 – move decimal ONE to RIGHT
x 100 – move decimal TWO to RIGHT

2.34 x 10 = 23.4
2.34 x 100 = 234

Multiply any number by 2

x 2– Double it   x 2% - Double it
                                  then – move the decimal TWO to LEFT

165 x 2 = 330   145 x 2% = 2.90

Multiply any number by 5

WHOLE Numbers

x 5 –   Step 1: add a zero   x 5% - move decimal ONE to LEFT
         Step 2: cut in half   cut in half

48 x 5   = 480/2 = 240   48 x 5% = 4.8/2 = 2.4
327 x 5 = 3270/2 = 1635   327 x 5% = 32.7/2 = 16.35

DECIMAL Numbers

x 5 –   Step 1: move decimal ONE to RIGHT
         Step 2: cut in half

3.25 x 5 = 32.5/2 = 16.25
48.3 x 5 = 483/2 = 241.5

This review of multiplying by 10, 2 and 5 is an exercise that adds the ability to compensate for percentages and decimal places. With the review complete you should be ready to mix all the processes up for everyday math situations.

Mix it Up!

What follows are a few problems that can be quickly solved using a combination of processes. Hope you enjoy and understand the process.

6 x 6 x 6 = _____   = (6 x 6) * 6
= 36 x 6
= (36 x 5) + (36 x 1)
= 360/2 (180) + 36 = 216

15% tip for $34.98 . . . round up to $35
= (35 x 10%) + (35 x 5%)
= 3.5 + 1.75 = $4.25

14 x 5672 = _____   = (5672 x 10) + (5672 x 2)*2
= 56720 + (11344 * 2)
= 56720 + 22688
    = (567 +226 = 793) then (20 + 88 = 108)
    = 793 + 1; append 08
= 79408

17 x 342 = _____   = (342 x 10) + (342 x 5) + (342 x 2)
= 3420 + 1710 + 684
= 5130 + 684 = 5814

18 x 657 = _____   = (657 x 10)*2 − (657 x 2)
= 13140 − 1314
    = (131 − 13) then append (40 − 14)
= 11826

     – or –

= (657 x 10) + ([(657 x 2) x 2] x 2)
= 6570 + ([1314 x 2] x 2)
= 6570 + (2628 x 2)
= 6570 + 5256
    = (65 + 52 = 117) then (70 + 56 = 126)
    = 117 + 1; append 26
= 11826

456 x 8 = _____   = [(456 x 2) * 2] x 2
= (912 * 2) x 2
= 1824 x 2 = 3648

Squaring any number ONE below or above a previous square:

$26 \times 26 = (25 \times 25) + (25 \times 2) + 1 = 625 + 50 + 1 = = > 676$

$34 \times 34 = (35 \times 35) - (35 \times 2) + 1 = 1225 - 70 + 1 = = > 1156$

$91 \times 91 = 90^2 + 180 + 1 = 8100 + 181 = = > 8281$

FORMULA:
$$(x + 1)^2 = (x + 1)(x + 1) = x^2 + 2x + 1$$
$$(x - 1)^2 = (x - 1)(x - 1) = x^2 - 2x + 1$$

Squaring any number TWO below or above a previous square:

$17 \times 17 = (15 \times 15) + (15 \times 4) + 4 = 225 + 60 + 4 = = > 284$

$33 \times 33 = 35^2 - (35 \times 4) + 4 = 1225 - 140 + 4 = = > 1089$

$62 \times 62 = 60^2 + 240 + 4 = 3600 + 244 = = > 3844$

FORMULA:
$$(x + 2)^2 = (x + 2)(x + 2) = x^2 + 4x + 4$$
$$(x - 2)^2 = (x - 2)(x - 2) = x^2 - 4x + 4$$

Squaring any number quickly between 10 and 120

All numbers can be found within two numbers of a 10 unit or a 5 unit; therefore, knowing how to square 1 or 2 above or below makes solving squares easy. Looking at the following numbers between 10 and 30 you can see that any given number is either a ten (10, 20, 30) or five number (15, 25)

10 ... 11 (+ 1) ... 12 (+ 2) ... 13 (− 2) ... 14 (− 1) ... 15
15 ... 16 (+ 1) ... 17 (+ 2) ... 18 (− 2) ... 19 (− 1) ... 20
20 ... 21 (+ 1) ... 22 (+ 2) ... 23 (− 2) ... 24 (− 1) ... 25
25 ... 26 (+ 1) ... 27 (+ 2) ... 28 (− 2) ... 29 (− 1) ... 30

You should remember how to square any ten unit number easily or any number ending in five. Knowing this and the formulas above for squaring any number one/two above or one/two below a number by 10 or unit of 5 you will become a squaring genius!

158

Even Divisibility by 11

You can determine if a number is evenly divisible by 11 by following these steps:

1. Breaking the digits apart
2. Adding all of the odd positioned digits together
3. Adding all the even positioned digits together
4. Subtracting the even from the odd digits
5. If the answer is divisible by 11 the number is also

EXAMPLE:

Is 39364831 evenly divisible by 11?

Add all ODD digits – 3 + 3 + 4 + 3 = 13

Add all EVEN digits – 9 + 6 + 8 + 1 = 24

Subtract one from the other

13 – 24 is same as 24 – 13 = 11   YES!

Another way is to change the sign of each digit and do the math in place:

$$39364831 - +3 - 9 + 3 - 6 + 4 - 8 + 3 - 1 = -11$$

Is 8563241 evenly divisible by 11? $+8 - 5 + 6 - 3 + 2 - 4 + 1 = 5$ NO!

Rapid Addition using 'Process of 11'

You can quickly add any series of numbers via a system known as 'Process of 11'. It is fairly easy to understand. Follow these steps to use the Process of 11

1. Add each column separately and mark off each instance of 11
   In the example below every time an 11 is reached an asterisks (*) is placed next to the digit
2. Place the remainder of addition per column on line one of solution
3. Place the number of units of 11 on the line immediately underneath plus the number of 11s from column to right
4. Add two columns to obtain the final solution (and any carry)

EXAMPLE:

78654	0	7	8	6	5	4	
281				2	8*	1	
5628			5*	6*	2	8*	
23154		2	3	1	5	4	
+ 22399	+	2*	2	3	9*	9*	
		0	7	7	7	4	(left over)
	+	1	2	2	3	4	2	(number of 11s plus neighbors)
		1	3	0	1	1	6	< < = = ANSWER

The Process of 11 is just another method you can use for rapid addition!

Endgame for the Day

Seeing is deceiving! Today's end game is more in-depth than those in the past. It discusses Optical Illusions and its science; showing different types.

What is an Optical Illusion?

A loose definition for an optical illusion is any image that is perceived by your eyesight to be visually deceptive.

There are many different divisions of optical illusion. Some related to distortion of shape, color, pictures, size and distance; others geometric in nature; even others affect sensory perception and hallucinogenic illusions.

Mathematics and Optical Illusions

All optical illusions occur within our minds. They are based upon a calculation of what is seen by our vision. The calculation is based on what the brain sees by "filling in the blanks" The brain performs these calculations, often geometric in nature, but not always – often by interpreting color or shading.

Natural Optical Illusions

When you use a pair of binoculars, you see things bigger/closer than they actually are. It is also true when using the outside rearview mirror of a vehicle – "Objects appear closer than they are!" Both examples are a form of an optical illusion. Another, when the moon is close to the horizon on an autumn night; it appears much larger than after it has risen in the sky.

The Brain Sees the Illusion

There are times that part of your brain tells you exactly what it is you are looking at; but when someone points out that the optical illusion is something specific (or has a specific trait), you suddenly see it! That "I see it now" observation is an "Ah Ha" moment when you believe you were tricked into seeing something that wasn't real. In reality, it is your brain that is doing all the math manipulation – even though it realizes it is messing up thus forcing you to look at it more carefully.

Ever wonder why people see the illusions? It is due to the human brain and how it is designed. The brain is designed to fill in blanks automatically, even though all data may not be immediately available. Ordinarily, this provides a more useful picture of what we think we see, by interpolating data to cover things we don't actually see.

The Eye and Brain connection

An optical illusion occurs when different cells and receptors of the eye perceive images and colors at different rates; resulting in a false image being sent to the brain.

When your eye "sees" anything, it does not actually see, or record, the information that it encounters; rather it is sent to the brain. In fact the retina, where the optic nerve is attached, has a blank spot. So if we saw what our eyes see, a blank spot would be in the middle of our field of vision. Instead, when the brain receives the visual content from our eyes, it fills in the blanks, and we see what appears to be a continuous image with no little missing spot.

Understanding this trait, allows exploitation of it. When we witness something, or remember seeing an incident, it is usually a "best guess" that we formulate based on memory. Thus it isn't always 100% accurate.

So, we see what our brains expect us to see (based on limited data); thus the ability to see what is not there – an optical illusion!

A few Optical Illusions

There are many illusions based on geometry. Here are a few:

The Müller-Lyer Illusion

Here is the *Müller-Lyer* illusion. It is an optical illusion based on the orientation of arrowheads on line segments.

Looking at the figure – are the two line segments equal in length? The top figure is the same as the bottom one (except the segments are separated.) Hmmm ....

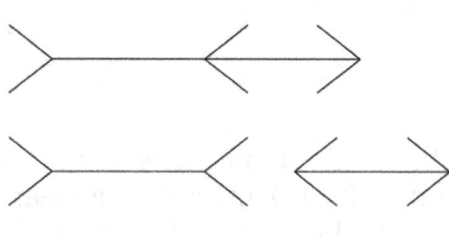

YES they are!

Vertical – Horizontal Illusion.

Here are two examples

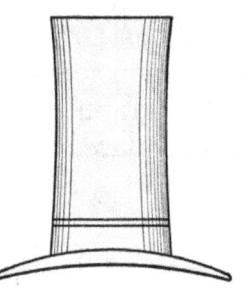

Vertical line segment appears to be longer than the horizontal segment. They are the same size.

To the right is another illusion based on the same process – the hat is NOT taller than it is wide.

Poggendorff Illusion

Another optical illusion is based on two ends of a straight line segment passing behind an obscuring rectangle. The two line segments appear to not be continuous. Looking at the illusion to the right; which line segment along the bottom of the rectangle is a continuation of the line segment across the top? Believe it or not – it is the bottom one.

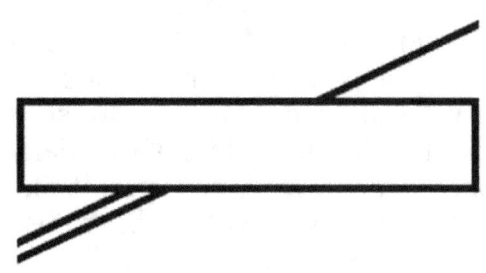

Zöllner Illusion

The next examples are based on a discovery by F. Zöllner in 1860 of parallel lines that appear to be at an angle. The one on the left was the one he discovered and sent in a letter to his physicist friend and scholar J. C. Poggendorff.

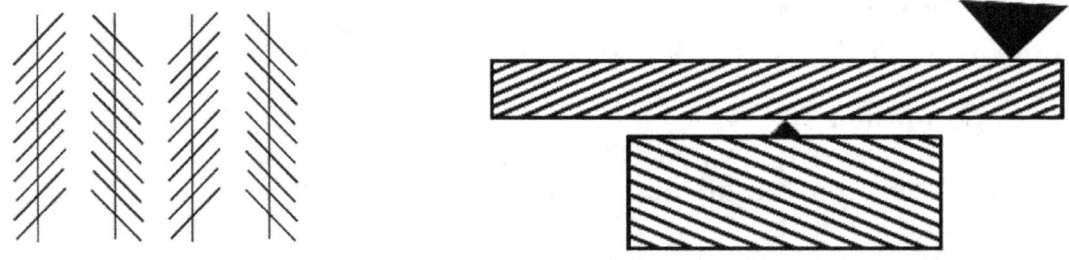

Both of these illusions demonstrate the visual illusion of vertical (left) or horizontal (right) lines that appear to not be parallel. The "scale" on the right appears to be balanced leaning downward on the right side. However, both rectangles are parallel.

Orbison's Illusion

This next illusion demonstrates how the mind can be fooled beyond simple parallel lines. This illusion, known as *Orbison's Illusion,* illustrates how multiple two dimensional objects can be deceptive.

The graphic shows how the bounding rectangle and inner square both appear distorted.

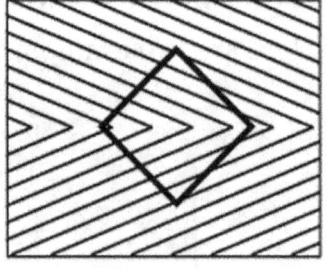

Fraser's Spiral

Two optical illusions named after British psychologist James Fraser are shown (1908). They are also known as a false spiral, or twisted cord illusion.

Both images appear to be spirals. On the left, the actual spiral studied by Frazer appears to be formed by a rope containing twisted strands of two different shades. It actually consists of concentric circles of twisted cords – no spirals there. The image on the right is actually four circles, one inside another.

Illusory Contour Figures

Here are radial line segments whose inward-end produce an illusion of a circle being present. The lines have the same color as the background; but appear brighter.

The contour illusion on left is also known as the Ehrenstein illusion.

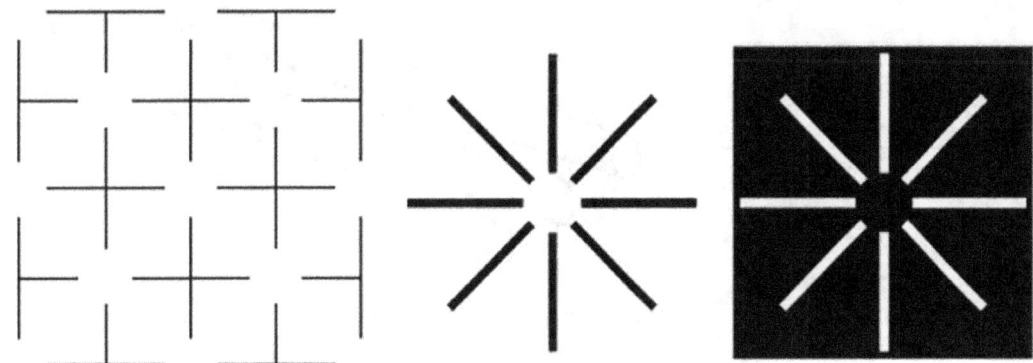

See the circles? They aren't really there!

Center-bar in a range of gray

Here is an interesting illusion. Is the center bar a range of gray or a single color?

Answer: Check for yourself –YES, the bar, is the same shade of gray throughout.

Impossible Shapes

Here are two impossible shapes – are there three or four shelves and is that trident real?

 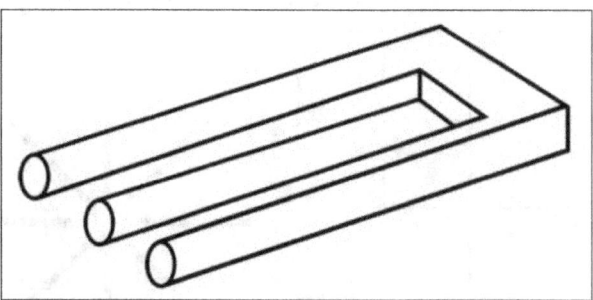

These optical illusions are good examples of how the eyes are fooled when we try to process three dimensional objects drawn in a two dimensional world.

Three dimensional world

Here is a picture of floor tiles in Basilica of St. John Lateran (Rome, Italy), which create an optical illusion. Hard to believe this is a floor – looks like steps.

Picture was taken by Tino Warinowski, June 2006, released to the public domain.

The Black Dot

Here are two different illusions that require a little bit of work to see their effect. Look at either dot and the surrounding rings (inner on left – outer on right) will disappear as you focus on the dot in the center.

These Optical Illusions are all related to mathematics and how we process objects.

Day Six

W ELCOME TO DAY 6 OF MATH 4 2-DAY! We are nearing the end of this journey. Today we continue through the fascinating world of mathematics and discuss some new topics. We even touch on the world of the 4th dimension. Need to review any material that has been covered over the past five days? If yes, please go back and review it. We will discuss topics like Möbius strips, math of exponentiation, and even how to do math the ancient Egyptian way.

Today's Outline

Quotations of the Day
Bit of Fun in Math
>Top 10 Excuses for not doing homework!
>There are 1 0 types of people in the world
Mathematical Curiosities – (Part 5)
>Möbius and his impossible shapes!
Exponentiation or Rising to the Power of
Calculator Tricks –(Part 6)
>General MacArthur's Favorite Number Game
Simple Math is All You Need – (Part 6)
>Convert Fahrenheit to Celsius and back
>Grab Bag of Methods
Endgame for the Day
>Mathematics and the ancient Egyptian way!

Objectives:

(1) Understanding the math of powers.
(2) Understand Möbius and his shapes
(4) Continue to perform mental math – avoiding the calculator
(3) Work with ancient Egyptian math

Quotations of the Day

ℰ · ℛ

If people do not believe that mathematics is simple, it is only because they do not realize how complicated life is.

-- John Louis von Neumann

John Louis von Neumann (1903-1957) was a Hungarian Professor of Mathematical Physics at Princeton University. He is best known for creating one of the fastest computers used by the US government to build and test the first model of the hydrogen bomb. He also developed a computer that performed twenty-four-hour weather predictions in a few minutes; through mathematical simulation of logistic problems. At the age of 34, he won the Bocher Prize (American Mathematical Society); he was a member of the National Academy of Sciences, American Philosophical Society, and American Academy of Arts and Sciences.

ℰ · ℛ

Mathematics is expected either to be immediately attractive to students on its own merits or to be accepted by students solely on the basis of the teacher's assurance that it will be helpful in later life. [Yet,] mathematics is the key to understanding and mastering our physical, social and biological worlds.

-- Morris Kline

Morris Kline (May 1, 1908 – June 10, 1992) was a Professor of Mathematics, New York University. During World War II, Kline helped develop RADAR. He wrote many books on various topics of mathematics. He believed in the need to teach the application and usefulness of mathematics rather than having students do math for the sake of learning it.

ℰ · ℛ

Small minds discuss persons. Average minds discuss events. Great minds discuss ideas. Really great minds discuss Mathematics!

Unknown Author

Bit of Fun in Math

Hopefully, you will enjoy these two topics presented for the last day of class. The first is about the 1 0 types of people in the world. Finally we finish with the top 10 excuses for not doing your homework.

There are 1 0 types of people in the world

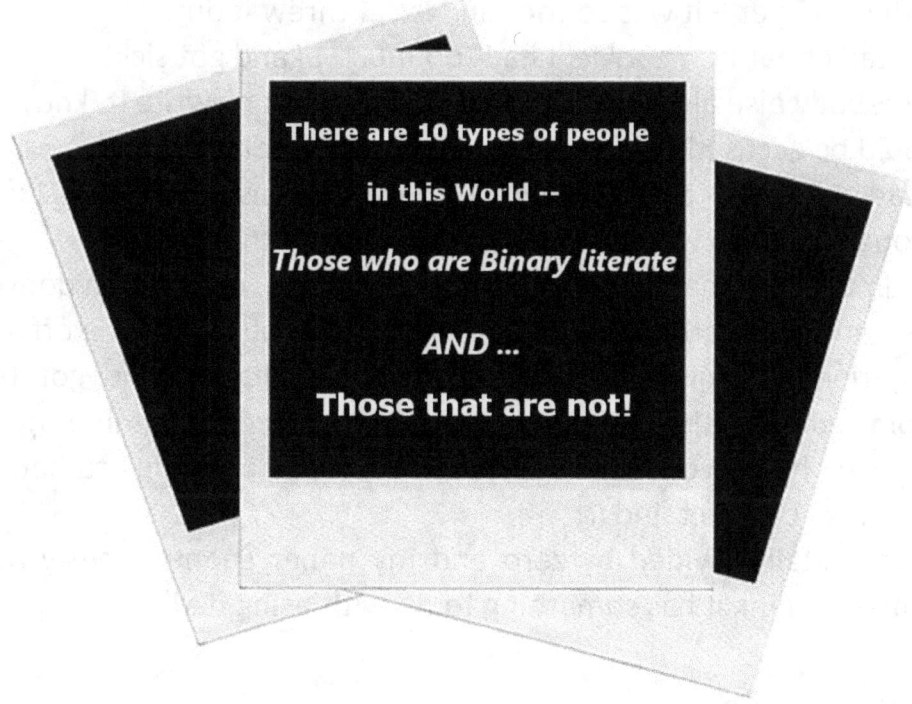

Knowing how computers work. This should make perfect sense. Yes?

So which type of person are you?

Top 10 excuses for not doing your homework!

1. I took time out to snack on a doughnut and a cup of coffee. The coffee had me so jittery that I couldn't hold my pencil.
2. I have a solar powered calculator; the electric was off last night.
3. I was re-reading my homework when I saw a spider and beat it to death with my paper – it was so torn and icky I threw it out.
4. Instead of eating a cookie, I had too much *pi* and got sick!
5. Somebody else already published it; so I didn't re-write it, knowing that I would be accused of plagiarism. I will give credit to them!
6. I had no paper and although I knew the answers, there just wasn't enough room to write them in the margins of the book.
7. I didn't realize that it was mandatory! Besides, my religion doesn't allow me to do any work after 3:00 p.m.; when the bus drops me off at home.
8. My friend promised she was going to do it; but she got busy and promised to finish it for me next week! (see excuse number 5)
9. I put the homework inside my Klein bottle before going to bed, but this morning I couldn't find it!
10. I accidentally divided by zero and my paper spontaneously burst into flames. I think it has something to do with using dad's "flash paper"!

In Closing:

Should you find that you really need additional help –

– it's just a phone call away!

Math problems?

Call:

$$\text{1-800- [INT}(20\pi)] \ (sin^2\theta + cos^2\theta)8 \ [0.1(7^2)] \ [\text{INT}(10e)+19(i)^4]$$

LEDGEND: $\pi = 3.14$ $I = \sqrt{-1}$ $e = 2.718$ $sin^2\vartheta + cos^2\vartheta = 1$

This number can actually be solved down to 1-800-MATH.WIN (using phone phonetics!)
NOT A REAL PHONE NUMBER!

Mathematical Curiosities – (Part 5)

The final mathematical curiosity we will explore is the study of objects that can fool the mind – no, not optical illusions; rather objects that are related to Möbius and his study of mathematical surfaces.

Möbius and his impossible shapes!

The *Möbius strip (Möbius band/loop)* is an object having a single surface; in mathematical terms it is said to have one boundary component (*topology*) with a property of being non-orientable. Non-orientable, in *topology*, refers to its having no defined top or bottom when created; rather a two-dimensional surface with a single side.

Like other discoveries in mathematics, the Möbius band was discovered independently by two German mathematicians, in 1858, August Ferdinand Möbius and Johann Benedict Listing. August was the first to demonstrate it in geometry and thus the strip was named after him.

> *Note*:
>
> Topology is a higher order of mathematics dealing with spatial properties that are maintained under processes of deformations, twisting, and stretching of an object. Topological ideas are present in many areas of today's mathematics It was actually defined by Johann B. Listing in 1836 in a letter he wrote to his old teacher Müller about his studies. He also wrote a book titled *Vorstudien zur Topologie* in 1847 that defined the mathematics of topology.
>
> A Topologist is a mathematician who studies qualitative questions about geometrical structures. Instead of asking how big something is … a topologist asks … does it have any holes in it? is it, as a whole, all connected together, or can it be separated into parts?

Creating a Möbius strip

A Möbius strip can be constructed by joining the two ends of a strip of paper after giving it a single half-twist. Once made, it will have a couple of unusual properties!

> *Note*:
> Directions for making a Möbius strip are on the next page.

First it only has a single surface; in other word no start – no end – no top – no bottom. Second, its' measured length is twice the length of the initial strip of paper.

CONSTRUCTING a Möbius strip

To create a Möbius strip start with an elongated rectangular piece of paper; put a single twist in it and finally, tape the two ends together. Here is a visual to help you:

```
┌─────────────────────────────────────────────┐
│ A                                          B │
└─────────────────────────────────────────────┘
```

Create a long rectangle (you can label each side if you like)

```
┌──────────────      ──────────────┐
│ A             ><              ᗺ │
└──────────────      ──────────────┘
```

Place a single twist in the rectangle (as shown)

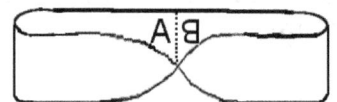

Tape the two ends together

If you created it properly, it will look like the one above with both sides taped.

Working with a Möbius strip

Once you create a Möbius strip there are certain things you can do with it.

It has a single surface. To demonstrate it, draw a line down the middle of the strip starting at any point. It will continue from the beginning and appear to wrap around the "other side" and will meet back at its starting point. This single continuous curve demonstrates that the Möbius strip has only one boundary in Topology terms.

If you measure the length of one side of the paper before you create the object and then measure the line you drew on the Möbius strip, it will be twice as long (double).

Is this possible? Well you just created one – you see it and feel it and it seems to defy the laws of Geometry.

Cut the strip along the line you created to show it only has one surface. Will it create two separate strips that are separate or linked together? Will it create one long strip with two twists in it? If you cut it, it will still be a single long object; containing two twists.

What will happen if you cut it down the middle again? Does it remain a single object with four twists in it; or become one very long object with a single twist in it? Perhaps two objects, tied together. It will be up to you to continue this experiment.

An example of the famous Möbius strip/loop is seen every day. It is the three-arrow symbol used to denote "recyclable". The Möbius strip is actually in public domain and can be used in any way.

Here is a gray scale representation of the original recycling logo, created by Gary Anderson, 1970.

Can you see the form of a Möbius loop/band?

Initially the plain Möbius loop, white with an outline or solid black, was to be used to indicate that a product was recyclable.

Eventually the Universal Recycling Symbol was made using a black outline and green fill.

A four dimensional (4D) Möbius object

The study of Topology has led to other three or more dimensional shapes. For instance, an interesting 4 dimensional object is known as a Klein bottle.

The Klein bottle is named after Felix Klein, who theoretically constructed it in 1882. He visualized two Möbius Loops sewed together to create a single sided bottle with no boundary. In other words, its inside is its outside. It literally contained itself.

If you recall, a Möbius strip is a half-twisted loop of a tape. If you were to make one large enough to walk on; you would walk along the loop, ultimately reaching the reverse side of the start point.

This simple structure is accomplished by interjecting a 3-dimensional-twist into the 2-dimensional world (a flat plane of a rectangle) of the paper: the band, that is created, is one-sided – having no start or end.

Here is how a Klein bottle can be formed from Möbius loops:

> Take a rectangle and join one pair of opposite sides. This will create a cylinder. Now join the opposite sides of the cylinder (other pair of sides) with a half-twist (like creating Möbius loop).

> UNDERSTAND, this last step is NOT possible in our universe – it requires 4-dimensions because the surface has to pass through itself without a hole.

A true Klein bottle can only exist in the fourth dimension. Yet every tiny slice of a Klein Bottle is 2-dimensional; therefore, a Klein Bottle is a 2-dimensional manifold which exists in the 4th dimension!

In mathematical terms you could say that a Klein bottle is a Riemannian manifold waiting to define a Euclidean metric at every point.

For those truly interested, here are a set of parametric equations that define the surface of every Klein Bottle:

x = cos(u)*(cos(u/2)*(√ 2+cos(v))+(sin(u/2)*sin(v)*cos(v)))
y = sin(u)*(cos(u/2)*(√ 2+cos(v))+(sin(u/2)*sin(v)*cos(v)))
z = -1*sin(u/2)*(√ 2+cos(v))+cos(u/2)*sin(v)*cos(v)

Here it is in polynomial form:

$$(x^2 + y^2 + z^2 + 2y - 1)[(x^2 + y^2 + z^2 + 2y - 1)^2 - 8z^2] + 16xz(x^2 + y^2 + z^2 - 2y - 1) = 0$$

Here are six steps necessary to create a Klein bottle:

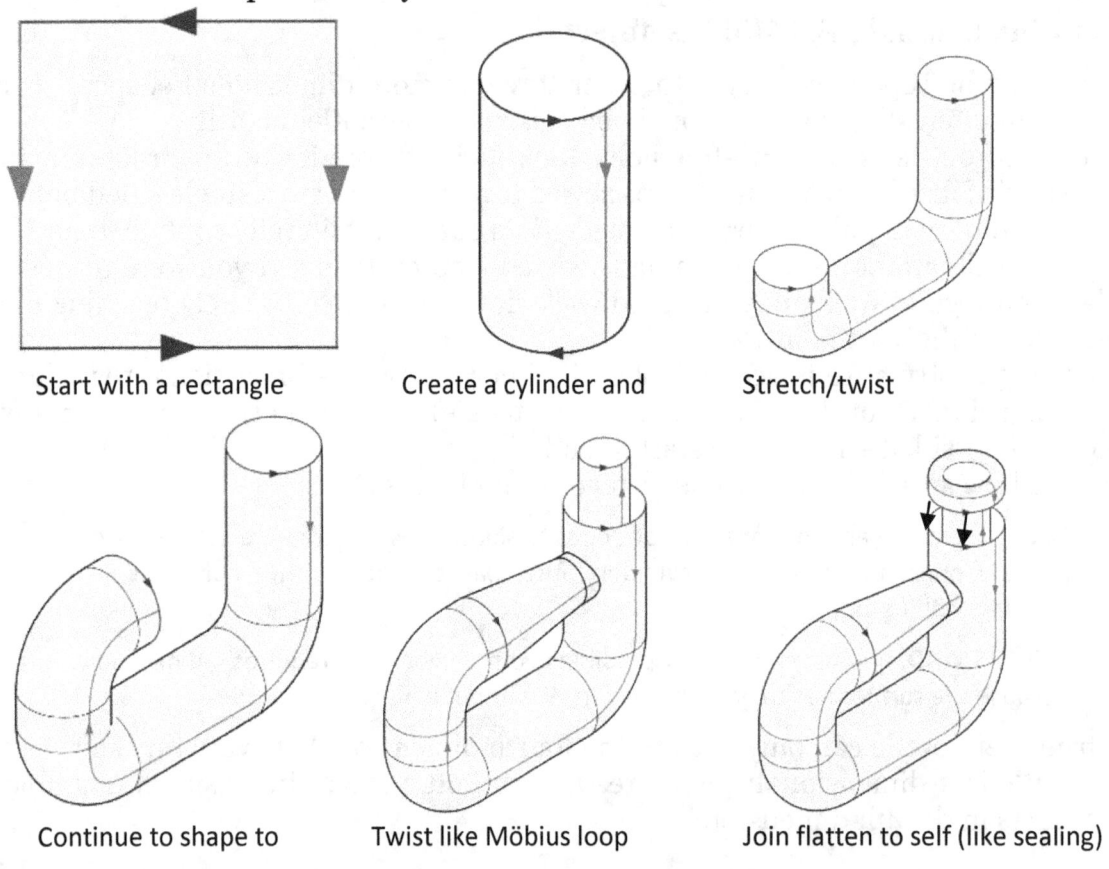

| Start with a rectangle | Create a cylinder and | Stretch/twist |

| Continue to shape to | Twist like Möbius loop | Join flatten to self (like sealing) |

– All drawings are in public domain –

<u>Note</u>: these last two steps CAN NOT be accomplished in the 3D world!

Understand that following these instructions would only create a pseudo-Klein Bottle. The real Klein bottle cannot be created in our 3D space.

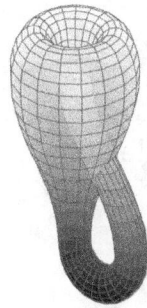

This is because of the step needed in the fifth drawing –in the real world you can't add the "4D-twist". So you can do something similar by imagining making a hole in the tube. Once made, bring the tube through upon itself and out. Then seam the opening of the 'hole' with the body of the tube. Finally, take the two separate ends and seal them upon each other.

The final Klein bottle would look something like this one, drawn with the program *gnuplot* version 4-4-2; (C) Copyright 1986 - 2009 Thomas Williams, Colin Kelley (FREEWARE).

The Klein bottle is a neat Topological concept. It has a single surface throughout so it is hard to visualize in the real world. The diagram on the previous page demonstrates how to create a Kline bottle.

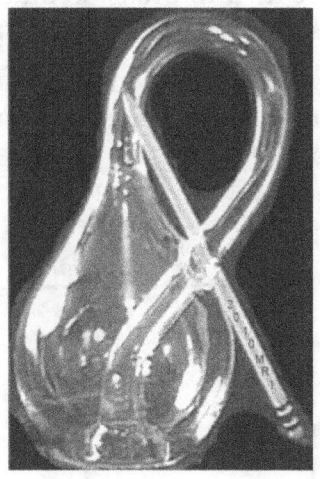

This is a picture of a Klein bottle that was created by Cliff Stoll, owner of ACME Klein Bottles in Oakland, California. You can learn more about these bottles (or even purchase one) at his website: kleinbottles.com.

Notice that there is one single opening located at the bottom of the picture.

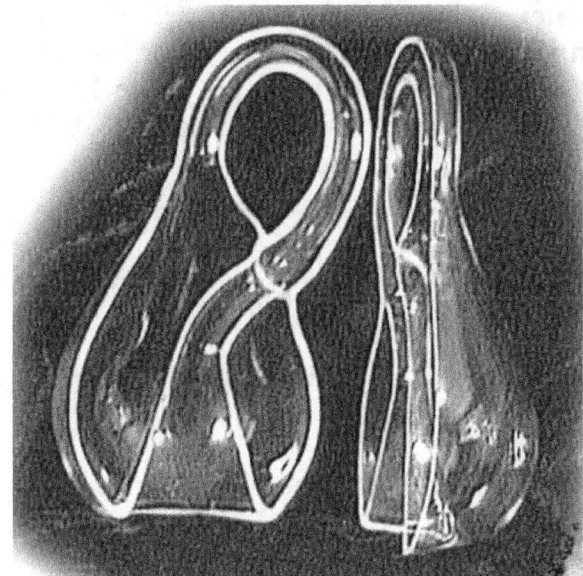

Closely examining the picture, to the left, this Klein bottle has been cut in half. You can see that the bottle has a single one-sided surface. If you take your finger and follow any point along the outside, you will soon realize it is also the inside.

It is just like the Möbius band/strip. In fact, each separate portion (small sliver) of the bottle would form Möbius bands.

WOW, a Klein bottle is really two Möbius bands with fill in-between!

A Möbius card

Here is another Möbius type object that was first demonstrated in the early 1930's. Most recently it was written about in the February 2002 *Games Magazine*. The article was titled *Möbius Card – An Impossible Shape*! By Chris Henderson.

The card is made from a 3 x 5 index card and appears to form an impossible shape. It appears to defy the laws of math with a single one-side surface.

Our mind tells us that this shape should not exist – but it does. More importantly, it is easy to create.

This shape may confuse some people when they look at it for the first time. Usually people will not be able to figure out how to create it until they physically make one and handle it. Here is how it is made:

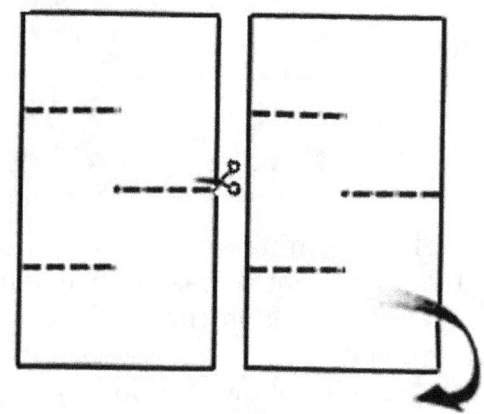

To create the object, mark your index card with three lines like those in the figure.

Then cut halfway into the card along each line. Do not cut beyond the half way mark.

Then, while holding the card with both hands on opposite corners and keeping it parallel to the floor; twist the lower half 180 degrees clockwise so that the top and bottom are now parallel to the floor.

Finally, crease the fold and put the tab up so it stands away from the remaining parts of the card.

Note:

When you twist the card to have the center part "pop" up, hold the card parallel to the floor and twist slowly – it will take shape.

This is a start into the fascinating world of Möbius and his impossible shapes.

Exponentiation or Raising to the Power of

Previously you worked with scientific notation and exponentiation, or "raising to the power of", when referring to a system of numeration. Remember 1.0×10^6 represents a million and 1.0×10^9, a billion?

What is Exponentiation

Exponentiation is a mathematical operation, written in the form x^n. It involves two numbers, the *base x* and the *exponent n*.

The number x^n can be read as: *x raised to the n-th power* or *x raised to the power [of] n*, or even more briefly: *x to the n-th*. Some exponents can be read in a certain unique way; for example x^2 is usually read as *x squared* and x^3 as *x cubed*.

The exponent simply says how many copies of the base are multiplied together.

Examples of positive powers of a number

To raise any *base x* number to a positive *exponent n* you multiply *n copies of x* multiplied together. Here are a few quick examples:

$$2^3 = 2*2*2 \ [8] \ \text{(3, 2s multiplied together)}$$
$$5^4 = 5*5*5*5 \ [625] \ \text{(4, 5s multiplied together)}$$
$$10^3 = 10*10*10 \ [1{,}000] \ \text{(3, 10s multiplied together)}$$

Another way of looking at it is:

2^2 (two to the power of 2) $\quad = 4 \ (2 * 2)$
4^4 (four to the power of 4) $\quad = 256 \ (4 * 4 * 4 * 4)$
7^3 (seven to the power of 3) $\quad = 343 \ (7 * 7 * 7)$
16^3 (sixteen to the power of 3) $= 4096 \ (16 * 16 * 16)$

Raising a number to a negative exponent (power)

If you raise a number to a negative power, you are really creating a fraction!

How can raising to a negative exponent (power) produce a fraction?

By definition, it just does. For example, 4^{-2} really means (converts) a fraction of $1/4^2$. This is an important concept in Algebra.

The opposite is also true, if the divisor is a number raised to a negative exponent, the answer becomes a non fraction positive number raised to a positive exponent.

Example:

$$4^{-2} = \frac{1}{4^2} = \frac{1}{16} \qquad\qquad \frac{1}{4^{-2}} = 4^2 = 16$$

Rules of Exponentiation and Algebra

Exponents are used in algebra problems, so it's important to understand the rules for working with exponents. Let's go over each rule in detail and see some examples.

Zero Rule

There is one "zero rule,"

<u>Any non-zero number raised to the power of zero equals 1.</u>

Note that **x** (as a non-zero number) can't equal 0.

$$x^0 = 1$$
$$x \neq 0$$

Rules of 1

There are two simple "rules of 1" to remember.

(1) <u>ANY number raised to the power of "one" equals itself.</u>
If multiplied 1 time, then it's logical that it equals itself.
(2) <u>1 (one) raised to any power is 1 (one).</u>
This is logical, because 1 x 1 x 1, as many times as you multiply it, is always equal to one.

$$x^1 = x$$

$$1^m = 1$$
$$1^4 = 1 \times 1 \times 1 = 1$$

Product Rule

The "product rule" says,

<u>When multiplying two powers that have the same base, you add the exponents together.</u>

In this example, you can see how it works. Just add the exponents together!

$$x^m \times x^n = x^{m+n}$$

$$4^2 \times 4^3 = 4 \times 4 \times 4 \times 4 \times 4$$
$$4^{2+3} = 4^5$$

Power Rule

The "power rule" says

<u>When you need to raise a power to a power, just multiply the exponents.</u>

So 5^2 raised to 3rd power equals 5^6!

$$(x^m)^n = x^{mn}$$

$$(5^2)^3 = 5^6$$

Negative Exponents

The "negative exponent rule" says

<u>Any nonzero number raised to a **negative** power equals its reciprocal raised to the compatible positive power.</u>

$$4^{-2} = \frac{1}{4^2} = \frac{1}{16}$$

Calculator Tricks –(Part 6)

Performing calculations that appear to be magical ... a few to try:

Superfast Addition

Start with a sheet of paper and,

1. Write down 2 numbers less than 20, one under the other
2. Make a third number by adding the first 2 together and write it below the first two
3. Then make a fourth number by adding the second and third, a fifth by adding the third and fourth, and so on, until there is a total of ten numbers
4. Now if you look at the numbers you can quickly give the total of all 10 numbers

> Answer: Multiplying the seventh number by 11 will give you the answer:
> Example 14+9+23+32+55+87+142+229+371+600 = 1562 || 142 x 11 1562 (1420+142)

Doubling three digits – again and again

PRESS Enter after every operation

1. Enter any three-digit integer
2. Multiply it by 11
3. Multiply the overall answer by 91
4. Now, check the result

> Answer: 11 x 91 = 1001, any 3 digit number times 1001 is the number twice!

MacArthur's magic 115

PRESS Enter after every operation

1. Enter Month of birth (Example February is 2, October is 10)
2. Times 2 (Double it)
3. Add 5
4. Multiply by 50
5. Add your age
6. Subtract 365
7. Now add 115 to the results

> Answer: The right most two digits are the person's Age.
> The remaining left digit/s is his Month of Birth

Note: This was an actual trick that General MacArthur would perform to amaze guests at parties he would attend.

Simple Math is All You Need – (Part 6)

Converting Fahrenheit to Celsius and back

There are traditional formulas to convert from °F to °C. All are based on knowing the freezing and boiling point of water in each scale – specifically 32°F or 0°C is the freezing point of water and 212°F or 100°C is the boiling point. Using these numbers you can create a scale like the one below:

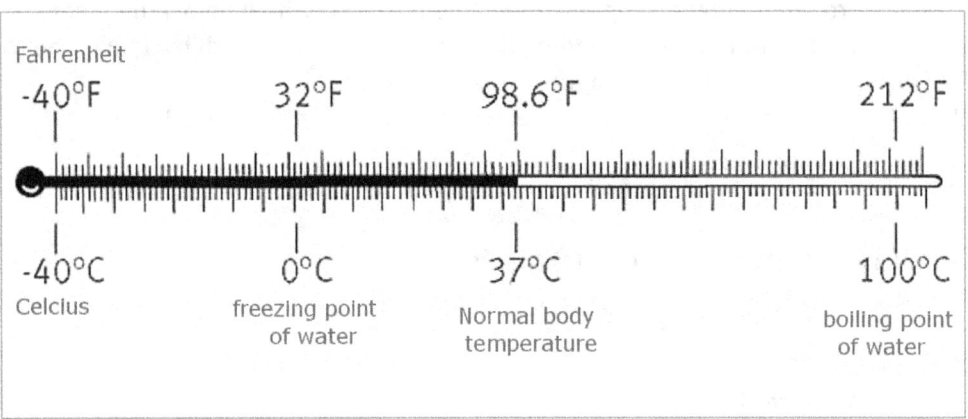

Understanding what the Scales say

Looking at the two scales, Celsius and Fahrenheit, certain relationships are clear.

- The degree difference in scales at the freezing point is 32 degrees. (base point 0)
- Difference in range from 32 to 212 –from freezing to boiling in Fahrenheit is 180 degrees.
- Different in range from 0 to 100 –from freezing to boiling in Celsius is 100 degrees.
- Relationship between the scales (moving from °F to °C) is 180-to-100 (90-to-50 or 9-to-5).
- Ratio of scales from °C to °F it is the opposite: 100-to-180 or 5-to-9.

Traditional Conversion Fahrenheit to Celsius

By making both values start at a common base point of their relative zero degrees, you can determine the ratio between each scale. It becomes obvious that for each increase of 9° Fahrenheit there is a corresponding rise in Celsius of 5°.

Understanding this correlation (increase in temperature from F°-to-C° is a ratio of 9° to 5°), you can determine the temperature in Celsius based on a temperature in Fahrenheit.

Steps to convert from Fahrenheit to Celsius:

1. Begin by subtracting 32° from the Fahrenheit number
   This starts the value at the Celsius base point (0)
2. Divide the answer by 9
   The number of degrees of change in relation to Fahrenheit
3. Then multiply that answer by 5
   The number of degrees of change in relation to Celsius

Here's an example:

It is a hot Ohio summer with the temperature already 98 degrees Fahrenheit. My friend, who lives in the Philippines, asks me what the temperature is. They use the Celsius scale. So I convert from F to C:

1. *98 degrees Fahrenheit* minus 32 gives 66.
2. Then, take 66 and divide it by 9 to get 7.33
3. Finally, 7.33 times 5 and you are given *36.65 degrees in Celsius*

Traditional Conversion Celsius to Fahrenheit

A similar process is used to convert from Celsius to Fahrenheit you simply follow these steps:

1. Begin by multiplying the Celsius temperature by 9
   This is the number of degrees to change in relation to Fahrenheit
2. Divide the answer by 5
   The number of degrees to change in relation to Celsius
3. Finally add 32
   This adjusts the base zero in Celsius back up to Fahrenheit

Here's an example:

Change a hot summer Manila 41 degrees Celsius to its corresponding degrees in Fahrenheit:

1. *41 degrees Celsius* times 9 and you get 369
2. Then, take 369 and divide it by 5 to get 73.8
3. Finally, take 73.8 and add 32 to get *105.8 degrees in Fahrenheit*

Algebraic Formulas for conversion

Remembering all of this can be a bit confusing. Here are the algebraic formula converting from Fahrenheit to Celsius and back:

$$T_C = (T_F - 32) * (5/9)$$

T_C = temperature in Celsius, T_F = temperature in Fahrenheit

To convert from Celsius to Fahrenheit, use this formula:

$$T_F = [(9* T_C)/5] + 32$$

T_F = temperature in Fahrenheit, T_C = temperature in Celsius

These formulas are better, yet there is *still* an even easier way.

The Fastest way to convert temperatures back and forth

To convert temperatures from Fahrenheit to Celsius or Celsius to Fahrenheit you need to remember three numbers:

40   **5**   **1.8** (from 9/5)

By keeping these numbers fresh on your mind, you are ready to learn a new method of conversion from one temperature scale to another.

Follow these steps to convert from *°F to °C*:

1. Add 40
2. Multiply by .55555 (decimal then *five* 5's)
3. Subtract 40

Follow these steps to convert from *°C to °F*:

1. Add 40
2. Multiply by 1.8
3. Subtract 40

The steps are very similar for each conversion. They have the same first and last step – add 40 at the beginning /subtract 40 at the end! Even the middle step uses the same operation – multiplying. The only difference is the middle step.

- *°F to °C* you multiply by 0.55555 (five 5's)
- *°C to °F* you multiply by 1.8

Add 40; multiply by a number (either .55555 or 1.8); then subtract 40.

MUCH EASIER and IT WORKS!

A Grab Bag of Methods

Squaring any number ONE below or above a previous square:

Knowing this formula you can quickly solve for any square that is one above or below a known square, like ending in 10 or 5.

PROOF: $$(X + 1)^2 = (X + 1)(X + 1) = X^2 + 2X + 1$$
$$(X - 1)^2 = (X - 1)(X - 1) = X^2 - 2X + 1$$

Squaring any number TWO below or above a previous square:

Knowing this formula you can quickly solve for any square that is two above or below a known square, like ending in 10 or 5.

PROOF: $$(X + 2)^2 = (X + 2)(X + 2) = X^2 + 4X + 4$$
$$(X - 2)^2 = (X - 2)(X - 2) = X^2 - 4X + 4$$

1/2 Fractions display 12's tables

Using a pattern based on showing equals of 1/2 you can do the 12 times tables: The first half of the fraction (top number) is the number being multiplied by 12. The bottom half of the fraction is determined by doubling (x 2) the top number.

Fraction	Equals	Times	Notes
1/2	= 12	x 1	Put 1 & 2 together
2/4	= 24	x 2	Put 2 & 4 together
3/6	= 36	x 3	Put 3 & 6 together
4/8	= 48	x 4	Put 4 & 8 together
5/10	= 60	x 5	Put 5+1 = 6 then 0 together
6/12	= 72	x 6	Put 6+1 = 7 then 2 together
7/14	= 84	x 7	Put 7+1 = 8 then 4 together
8/16	= 96	x 8	Put 8+1 = 9 then 6 together
9/18	= 108	x 9	Put 9+1 = 10 then 8 together
10/20	= 120	x 10	Put 10+2 = 12 then 0 together
11/22	=132	x 11	Put 11+2 = 13 then 2 together
12/24	=144	x 12	Put 12+2=14 then 4 together

... and so on!

Hope you find this a handy little trick for your twelve tables!

Rapid Multiply two digits different by 2

You can quickly multiply any 2, two digit numbers, separated by two numbers:

e.g. 24 x 22   34 x 36   73 x 71   11 x 13 and so on

1. take the number in between and square it (times itself)
2. subtract 1 from the resulting square
3. write the answer

EXAMPLE:

$$22 \times 24 = \underline{\hspace{1.5cm}}$$
$$23^2 - 1 =$$
$$(25^2 - 100 + 4) - 1=$$
$$= 528$$

$$34 \times 36 = \underline{\hspace{1.5cm}}$$
$$35^2 - 1 =$$
$$1225 - 1 =$$
$$= 1224$$

Rapid Multiply Two Numbers having a special relationship

RELATIONSHIP: The Tens digits are the same and Units digits add to 10

e.g. 24 x 26   34 x 36   71 x 79   52 x 58 and so on

1. Multiply the tens unit digit by the next higher digit (left most digit/s)
2. Multiply the unit's digits by each other (right two digits)
3. Combine the numbers

EXAMPLE:

$$26 \times 24 = \underline{\hspace{1cm}}$$
$$(2 \times 3) = 6$$
$$(6 \times 4) = 24$$
$$= 624$$

$$71 \times 79 = \underline{\hspace{1cm}} \quad 52 \times 58 = \underline{\hspace{1cm}}$$
$$(7 \times 8) = 56 \qquad (5 \times 6) = 30$$
$$(1 \times 9) = 9 \qquad (2 \times 8) = 16$$
$$= 5609 \qquad = 3016$$

Rapid Multiply Two digits times 101

e.g. 24 x 101   -   71 x 101   -   52 x 101 and so on

1. Write the two digits down twice

EXAMPLE:

$$24 \times 101 = 2424$$
$$72 \times 101 = 7272$$
$$87 \times 101 = 8787$$
$$52 \times 101 = 5252$$

Rapid Multiply Three digits times 101

e.g. 124 x 101   or   456 x 101, and so on

1. Write the two left most digits on left of the number
2. Write the two right most digits on the right (leave blank in middle)
3. Add left most and right most digit and put in middle

<u>EXAMPLE</u>:

$$124 \times 101 = \underline{\qquad}$$
$$12 \underline{\quad} 24$$
$$1 + 4 = 5$$
$$12\,\underline{5}\,24$$
$$= 12524$$

$$456 \times 101 = \underline{\qquad}$$
$$45 \underline{\quad} 56$$
$$4 + 6 = 10$$
$$1$$
$$45\,\underline{0}\,56$$
$$= 46056$$

Another way to square any number

This is just another way to square any number. It works very efficiently.
It requires these steps:

1. Find the closest number that ends in 0 (10s 100s etc.)
2. Write the tens number above and how far it is away (number to get to tens number)
3. Add the distance number to the original number to be squared, place it below and the distance number again.
4. Multiply the tens number times the number below the original
5. Add the square of the distance number to the sum from step 4

Here are three numbers to show how it works; e.g. 13^2 76^2 193^2 and so on

$$13^2 \begin{array}{l} {\scriptstyle -3}\ \ 10 \\ {\scriptstyle +3}\ \ 16 \end{array} \quad \begin{array}{l} = 160 + 9 \\ = 169 \end{array} \qquad 76^2 \begin{array}{l} {\scriptstyle +4}\ \ 80 \\ {\scriptstyle -4}\ \ 72 \end{array} \quad \begin{array}{l} = 5760 + 16 \\ = 5776 \end{array}$$

$$193^2 \begin{array}{l} {\scriptstyle +7}\ \ 200 \\ {\scriptstyle -7}\ \ 186 \end{array} \quad \begin{array}{l} = 37200 + 49 \\ = 37249 \end{array}$$

13^2 is 3 (distance) from 10. (1) Place the distance and 10 above. (2) Then add the distance (3) to 13 to get 16 and place 16 below. (3) Then multiply 10 x 16 for 160 (4) take the square of 3, *9*, and add it to 160 +9 = 169 or 13^2!

Fun with numbers and oddities

NUMBERS THAT MAKE YOU GO .... AHHHH!
Here are a few numbers that offer interesting relationships:

$12 \times 12 = 144$	: is to :	$21 \times 21 = 441$
$13 \times 13 = 169$	: is to :	$31 \times 31 = 961$
$12 \times 22 = 264$	: is to :	$22 \times 21 = 462$
$102 \times 102 = 10404$	: is to :	$201 \times 201 = 40401$

Did you consider these?

$$81 = (8 + 1)^2 = 9^2$$
$$512 = (5 + 1 + 2)^3 = 8^3$$

*Here are a couple of NARCISSISTIC numbers ***

$$153 = 1^3 + 5^3 + 3^3$$
$$371 = 3^3 + 7^3 + 1^3$$

*** a narcissistic number is one that is created from the sum of each digit raised to the power of the number of digits

Raising by the summation and power of sequence

$$13^5 = 1^1 + 3^2 + 5^3$$
$$17^5 = 1^1 + 7^2 + 5^3$$

Digital delight of numbers (1 – 9)
For these problems, you need to create valid addition problems by using all nine digits (1 through 9) once and only once –

243	341	317	154
+675	+586	+628	+782
918	927	945	936

There are other solutions ...

<u>Note</u>:
Did you notice that each solution begins with a 9 and is the sum of the other two digits?

Two's company

An interesting problem: Using five 2s and four basic operators (+, −, x, /) you can create a series of problems that have an answer equal to the digits 0 through 10.

$2 - (2/2) - (2/2) = 0$	$2 \times 2 \times 2 - 2 - 2 = 4$	$2 \times 2 \times 2 + 2 - 2 = 8$
$2 + 2 - 2 - (2/2) = 1$	$2 + 2 + 2 - (2/2) = 5$	$(2 \times 2 \times 2) + 2/2 = 9$
$2 + 2 + 2 - 2 - 2 = 2$	$2 + 2 + 2 + 2 - 2 = 6$	$2 + 2 + 2 + 2 + 2 = 10$
$2 + 2 - 2 + (2/2) = 3$	$(22/2) - 2 - 2 = 7$	$(22 \times 2)/(2 + 2) = 11$

You can use other operators to come up with other numbers also. Try this same problem using 4's instead of 2s!

The beauty of symmetrical math

A few exciting series follows:

$1 \times 8 + 1 = 9$	$1 \times 9 + 2 = 11$
$12 \times 8 + 2 = 98$	$12 \times 9 + 3 = 111$
$123 \times 8 + 3 = 987$	$123 \times 9 + 4 = 1111$
$1234 \times 8 + 4 = 9876$	$1234 \times 9 + 5 = 11111$
$12345 \times 8 + 5 = 98765$	$12345 \times 9 + 6 = 111111$
$123456 \times 8 + 6 = 987654$	$123456 \times 9 + 7 = 1111111$
$1234567 \times 8 + 7 = 9876543$	$1234567 \times 9 + 8 = 11111111$
$12345678 \times 8 + 8 = 98765432$	$12345678 \times 9 + 9 = 111111111$
$123456789 \times 8 + 9 = 987654321$	$123456789 \times 9 + 10 = 1111111111$

Here is one more example of mind boggling symmetry in mathematics:

$$1 \times 1 = 1$$
$$11 \times 11 = 121$$
$$111 \times 111 = 12321$$
$$1111 \times 1111 = 1234321$$
$$11111 \times 11111 = 123454321$$
$$111111 \times 111111 = 12345654321$$
$$1111111 \times 1111111 = 1234567654321$$
$$11111111 \times 11111111 = 123456787654321$$
$$111111111 \times 111111111 = 12345678987654321$$

Endgame for the Day

Have fun with this one – the math of ancient Egypt.

Doing Ancient Egyptian Math

Early Egyptians performed math using hieroglyphic symbols that represented their numbers. The symbols they used were:

Number	Symbol	English name
1		Single stroke
10		Cattle Hobble (or arch)
100		Rope Coil
1,000		Lotus Plant
10,000		Finger
100,000		Frog (or tadpole)
1,000,000		God Arms raised above head

Using these symbols you can interpret the value of any number and solve almost any mathematical problem found in hieroglyphics.

Hieroglyphic numbers are based on units of 10, something like our system for whole numbers. The difference is their physical numbers and ours is that theirs are pictures that are repeated a number of times for each of these units (1, 10, 100, 1,000, 10,000, 100,000, 1,000,000).

For example in our math we would write the numbers from left to right each position representing a unit of 10.

2,346 is really 2,000 + 300 + 40 + 6

In hieroglyphics numbers it is written :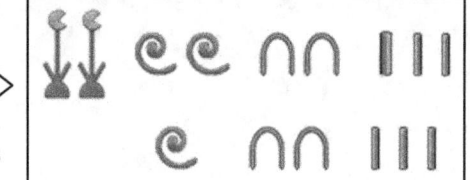

The number uses two lotus flowers for 2,000, three coils of rope for 300, four cattle hobbles (arch) for 40, and six stokes for 6.

Once you accumulate ten symbols (same); you replace them with one from the next group. For instance,

Ten cattle hobbles (arches) are equal to one rope coil. You would not write 10 hobbles to represent 100; rather replace all 10 hobbles with a coil for 100.

Using these simple rules, you can do simple math using hieroglyphics.

For example adding and subtracting is easy. To add two numbers you convert the symbols into numbers and then do the math and write your answer in hieroglyphics.

Problem: Add 1,009 and 3,202 in hieroglyphics:

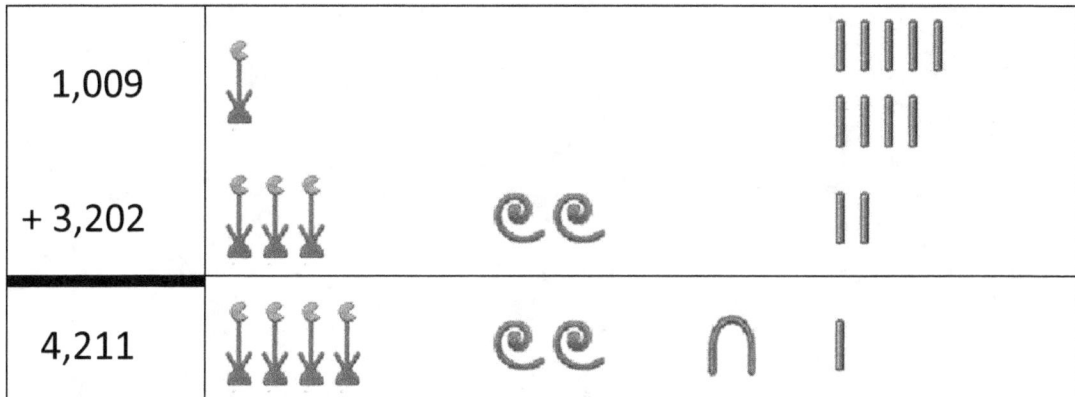

You do the math just like modern day math. First add the units (11) – carry 10 and put the stroke down for 1. Then add the tens with the carry getting 10 and put the arch down for 10. Then add the hundreds which total two hundred and put the two coils down next, finally add the thousands and put four lotus plants down.

Now you too can do math problems the hieroglyphic way!

I hope you enjoyed these six Days of Math 4 2-Day! Don't let the end of this course end your journey. I hope you will continue your trip through the world of Mathematics.

Michael R. Irwin

Appendix

Welcome to the Appendices for Math 4 2-Day! Here are a few additional topics that didn't formally make it into the course. It is due to the nature of a lecture or course. These topics are complimentary and deserve inclusion.

Author Guide and Comments

Any class is limited by something known as time. The amount of information that can be shared over any given class is limited to the time available. Remember, this book is centered on my Math-A-Nation course and covers the topics as presented each day. The difference between this book and the six days of classes is the depth of coverage for each topic. Unlike the actual course which is based on a two-hour window of time, the book can cover each topic in greater detail.

Yet there were a couple of topics that truly deserve to be in this book. The topics covered in this appendix are those 'fill-in' topics that come up from class to class based on questions and free time. Students will often raise a question that is loosely related to the subject being presented. For instance, "*I seem to have problems when doing subtraction. Is there an easier way to do it?*" or "*I have this problem that just doesn't make sense – it is an algebra problem.*"

This appendix has several problems. One is a method of doing subtraction that includes addition. Another is a simple graphic and question about a school bus. One is a simple algebra problem that seems to defy logic. Finally there are a few problems that require use of your brain power with the ability to observe.

So let us explore these additional topics:

Subtraction through Addition

People tend to have little or no problem when doing large number addition. Subtraction is another issue. For example many people seem to struggle with the subtraction problem like 7251 − 5454!

However, there is a method of performing subtraction that can take the fear away – it uses *addition to do subtraction*.

To understand how it works there is one simple rule to consider:

NEVER subtract a single digit from a number greater than 10

The best way to explain this rule is through a demonstration of the traditional subtraction method, using the problem below for illustration:

$$\begin{array}{r} 4\ 3\ 2 \\ -\ 2\ 7\ 9 \\ \hline \end{array}$$

To perform the arithmetic you need to start with the units and subtract 9 from 2. What you can't subtract 9 from 2 because it is bigger than 2? OK, then you need to borrow (take) 10 (1) from the tens column (30) subtracting 1 from the 3 and the problem now looks like this:

$$\begin{array}{r} 4^2\mathbf{3}^1 2 \\ -\ 2\ 7\ \ 9 \\ \hline \end{array}$$

Now you have changed two numbers on the top line – 2 now becomes 12 and 3 now becomes 2. Once this is done, you take 9 from 12 and get 3 (writing 3 below the units of the problem.)

Continuing, you try to take 7 from 2 and that can't be done; so you take (borrow) 1 (100) from the hundreds column and make the 2 now 12, as shown below:

$$\begin{array}{r} ^3 4^{12}\mathbf{3}^1 2 \\ -\ \ 2\ 7\ \ 9 \\ \hline 3 \end{array}$$

Now take 7 from 12 and place 5 under the second column. Finally take the remaining 2 from 3 and the solution looks like this:

$$\begin{array}{r} ^3 4^{12}\mathbf{3}^1 2 \\ -\ \ 2\ 7\ \ 9 \\ \hline 1\ 5\ 3 \end{array}$$

And the answer is 153! Notice all that borrowing (taking 1 from the previous column) and how messy the problem can become when striking out a number (putting a number one lower) when subtracting?

Thus the rule:

"*Never subtract a single number from a number higher than 10.*"

Using Addition while Subtracting

It is possible to never subtract from a number greater than 10 and supplement it with additional addition – all while doing subtraction!

It requires understanding a few concepts.

First: If the *top* number *is larger than* the *bottom* number,
            Subtract the bottom from the top; like in normal subtraction.

Second: If the *top* number *is smaller than* the *bottom* number;
            Place a 1 next to the number to its left (10's 100's, and so on)
            Subtract the bottom number from 10 and then
            add the top number to the result

With these concepts in mine here are the formal steps for doing subtraction math through addition:

1. If the top number is larger than the bottom – do the subtraction

2. If the top number is smaller than the bottom;

    (a) Subtract the bottom number from 10

    (b) Add the top number to the solution from step 2.(a)

    (c) Add a 1 to the next digit (column, left) of the current digit (bottom line)

    (d) Continue doing subtraction with next (left) column

Two examples demonstrate this process:

(a)	$\begin{array}{r} 19 \\ -\ \ 7 \\ \hline 12 \end{array}$	<<= = = 9 is larger than 7 so subtract normal Next Column is 1 – 0 also a smaller bottom number
(b)	$\begin{array}{r} 3\ 4 \\ -\ 1_16 \\ \hline 1\ 8 \end{array}$	<<= = = 4 is *less* than 6, so subtract 6 from 10, add 4 (top); 4+4=8 (write under first column). Add 1 to next column (bottom left) Do next subtraction: take 1+1 (2) from 3 and write 1

Looking at example (a) *19 – 7*, we subtract starting with the unit part (9 – 7). Since 7 is less than 9, just subtract like you normally would.

Example (b) *34 − 16*, on the previous page, is different. Starting with the unit part (*4 − 6*); 6 is less than 4 ... so ... instead of taking 1 (actually 10) from the top number in the next column; subtract *6 from 10* (4) and add the top number (4) to that answer (*4 + 4*) and then write answer under the first column.

NOW, place a small 1 next to the 1 in the next (left) column (tens).

Moving to the next column (3 -1), notice the small 1 next to the bottom 1 – this means that you need to add 1+1 together (2) before doing the next subtraction! This makes the tens column subtraction 3 -2 or 1. Placing the 1 under the column, the answer is 18.

Now on to a large subtraction problem:

$$7\ 2\ 5\ 1$$
$$-\ 5_14_15_14$$
$$1\ 7\ 9\ 7$$

STEPS:
1. 4 > 1; 10 - 4 (6) plus 1 = 7; put 7 below & add 1 to 5 (next column)
2. 5 +1 (carry) = 6; 6 > 5; 10 − 6 (4) plus 5 = 9; put 9 below & add 1 to 4
3. 4 + 1 (carry) = 5; 5 > 2; 10 − 5 (5) plus 2 = 7; put 7 below & add 1 to 5
4. 5 + 1 (carry) = 6; 6 < 7; 7 − 6 = 1; put 1 underneath – DONE!

Just remember, whenever the top number is smaller than the bottom number, subtract the bottom from 10, *then* add the top number, *and* carry a one to the next number on the bottom (left). Continue your subtraction until you are done.

Now it is your turn to do a few subtraction problems through addition.

3 1 9 4	7 3 7 4	3 4 2 7	2 8 9 4
− 1₁5 2₁7	− 4 4 3 6	− 2 2 8 6	− 1 9 9 7
1 6 6 7			

– Answers below –

So, there you have it – ***Subtraction using Addition*** !

Answers: (Subtraction using Addition)

7 37 4	34 27	2 8 9 4
− 4₁43₁6	− 22₁86	− 1₁9₁9₁7
2 9 3 8	11 41	8 9 7

194

School Bus Problem – What Direction

This is an interesting problem. It is extremely simple once you know the answer. Yet it stumps many people when they first encounter it.

Which way is the bus traveling, *LEFT* or *RIGHT*?

<u>*Note*</u>: Most Kindergarten children solve this problem in less than 20 seconds.

- Answer on Next Page –

= = = = = = = = = = = = = = = = = = =

Hint # 1 You live in America

= = = = = = = = = = = = = = = = = = =

Hint # 2 Where is the driver?

= = = = = = = = = = = = = = = = = = =

Hint #3 Where is the door?

ANSWERS (Which Direction is the Bus Going):

The bus is traveling to the LEFT ...

If it was going to the right you would see the door. Since the door is not visible, it must be on the other side of the bus. The door is also at the front of the bus. The driver sits across from the front door. Thus the driver is on the left of the bus and it is traveling LEFT.

Two Neurological Tests

There are many tests for intelligence and logic. The following aren't official, and some people even consider them jokes, yet I assure you they are not jokes. They are actually a good measure of observation skills.

A Test of 'F's

This test seems too obvious. It simply asks you to read a very short passage and answer a single question.

It is one of several email attachments from the "stranger and stranger" by Robert McMullen.

Read this sentence:

FINISHED FILES ARE THE RESULT
OF YEARS OF SCIENTIFIC
STUDY COMBINED WITH
THE EXPERIENCE OF YEARS.

Count the "F"s in the sentence. *Count them ONLY ONCE.* Do not go back.

How many "F"s did you count?

- Answer on Next Page –

Finding the 6 in a bed of 9s

999999999999999999999999999999999999999999999
999999999999999999999999999999999999999999999
999999999999999999999999999999999999999999999
699999999999999999999999999999999999999999999
999999999999999999999999999999999999999999999
999999999999999999999999999999999999999999999

- Answer on Next Page –

ANSWERS:

(A test of 'F's)

Did you answer 6? There are six Fs in the sentence. A person of average intelligence finds three of them. If you spotted four, you're above average. If you got five, you can turn your nose up at almost anybody. If you caught six, you are a genius. There is no catch. Many people forget the OFs. The human brain tends to see them as Vs not Fs.

(Find the '6' in a Bed of '9's)

Did you find it in the fourth row? Well that is where it is .... Fourth row, left side, first character.

Thinking Algebraically

Although we covered a few problems in Algebra during the past six days, I would like to discuss a problem that seems to defy logic. That is until you think about it algebraically.

Cost of a Baseball and Glove

This problem tends to stump many people. Upon first reading, the answer seems obvious and most people blurt it out – incorrectly. The answer that initially comes to mind is often *wrong*!

Read the problem completely before answering. Think it out logically and then write the algebraic formula that is used to solve it.

> While in the "Sports-4-All" store, you find a new baseball glove that you just have to have. You also decide to buy one baseball to use with the glove. You take them to the counter. The cashier rings up both items and announces the total cost will be $10.00. This store does not charge sales tax separately; the tax is included in the actual price for each item.
> You pay the person exactly $10.00, pick up your new glove and ball and leave the store.
> When you get home you look at the sales receipt and announce to yourself, "How about that, the glove cost exactly $9.00 more than the baseball."
>
> PROBLEM: How much did the baseball cost?

Take your time, write it out as an algebraic formula; making the baseball an unknown value (x). Then define the cost of the glove relative to (x). Write the formula.

Was your answer $1.00 for the cost of the baseball? If YES, you are wrong.

Still having problems figuring it out? Here are two hints:

1. The baseball is represented as variable *n*
2. The glove is *exactly* $9.00 *more than* the baseball

- Answer on Next Page –

197

> ANSWER: **(Cost of a Baseball and Glove)**
>   Did you solve it? The baseball cost $0.50 and the glove cost $9.50.
>   The formula would be
>   Cost of baseball = x
>   Cost of glove = x + 9
>   Formula is − (x) + (x + 9) = 10
>   2x +9 = 10
>   2x = 10 − 9
>   x = 1/2
>   x = $0.50 (fifty cents)

M, Heart, 8 Problem

If you ever watch the 2008 movie, *"The Oxford Murders"* by the Argentine mathematician and writer Guillermo Martínez, you will see this problem.

This Spanish-English-French production is an exciting murder thriller that takes place in England. It involves an Oxford University logic professor and an American student who team up to stop a series of murders.

The professor, played by John Hurt, offers some math, physics, and philosophical problems through-out the movie. At one point, while talking with his student and a police person, he draws a problem (here on the right) on the board making a general comment about how any true mathematician could easily solve it.

Now it is your turn. Solve the problem of what is the next symbol after the *"M"*, Heart, line, 8?

Ask yourself; is it a mathematical problem with some operation between the M and heart? Is the answer 8?

I can assure you that it is both a brilliant and easy problem to solve. It isn't an operation (like addition or subtraction). It is a logical problem that can have many more than one symbol after it.

Here is a Hint:

1. It is clearly some sort of symbolic symmetry
2. It is as simple as *1 − 2 − 3*

Give up?

– Answer on Next Page –

ANSWER: (**M, heart, 8**)
> Did you solve it? The next character that follows is:

-M-

Confused? See the explanation below.

The sequence of shapes in the original problem is just a series of mirror images of the number figures one, two, and three. So the next answer is the mirror image of the number 4.

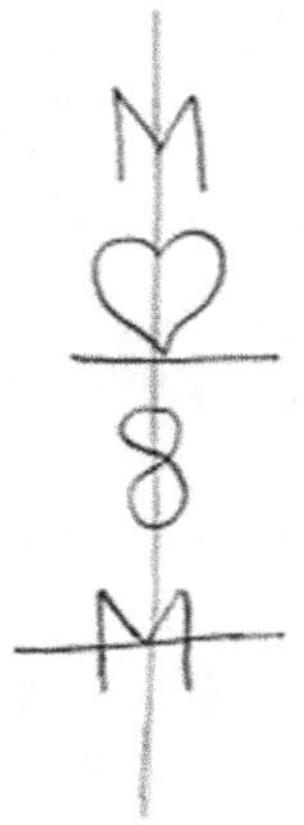

Here is the answer for the original problem, with a line drawn down the center. Notice how the line cuts through the middle of each symmetrical shape.

Looking closely you should see that each symbol has a 'mirror' image opposite it.

Do you see the numbers 1, 2, 3, and 4?

By drawing a line of symmetry down through the center of each symbol, the solution is more obvious.

Once you see the answer, it is pretty amazing and so simple to solve. The "M" becomes mirror ones; the Heart & line become mirror twos; the eight is mirror threes and of course the "M" with a line just below the center point becomes mirror fours!

Now I give you a challenge to solve. The above problem shows the next symbol. You can create five more. What are the next several shapes?

Hint: They will represent the numbers 5, 6, 7, 8, and 9?

SOLUTION:

5 will take on the look of a line across the top with a downward line from its center connecting to an oval. How about the 6? Well it will look like a pair of open scissors pointing upward. The 7, well a long staple should do it. Of course the 8 would be four circles two alongside each other and atop two others. Finally, since 9 is similar to 6, you can assume it will be another pair of open scissors pointed downward, lacking the upper blades.

- Answer on Next Page –

I will leave it to you to cover half of the picture and see the numbers 1 through 9.

INDEX

1- 5

12's Tables
  Easy method to perform, 183
196 Algorithm, 39
40, 5, 1.8
  Fast Temperature Conversion, 182
5 lines of 10 balls, 119

A

A Test of 'F's, 196
Addition Problems
  Check Digits, 12
  Using 'Process of 11', 159
addition to solve subtraction, 31
Ahmes
  Egyptian scribe, 108
Algebraic Numbers, 9
Al'Khwarizmi, 147
Ancient Egyptian Math, 188
Anderson, Gary, 173
Apocalyptic 666, 43
Archimedes, 94
  Principle of buoyancy, 95
  The Sand Reckoner, 95, 96
ASCII Character Set
  Computers, 151
Associative Property, 58
Avogadro number (constant), 99
Avoiding Careless Error Math, 10, 35, 138

B

Babbage, Charles, 56
Bacon, Roger, 28, 148
Base 10 System
  Convert to Base 2, 148
  Hindu-Arabic System, 147
Base 2 System, 147
  Binary & Computers, 148
  Convert to Base 10, 148
Base-1
  Unary System, 145
Base-10 System, 145

Bathers
  Golden Rectangles & Ratio, 137
Big Numbers. *See* Enormous numbers
Billion a Billion?, 43
Bit of Fun in Math, 3, 29, 57, 93, 123, **169**
Bits and Pieces of Algebra, 24
Black Dots, 165
Bonaccio, Guglielmo, 125
Brahmagupta, 108
  define Zero, 146
Breaking apart problems, 32
Brother Juan Diez Method Multiplication, 108, 113

C

Calculator Tricks
  4-2-1 Loop, 153
  6174 Loop, 115
  Birthday Display (2), 115
  Cheat the Calculator, 23
  Doubling three digits – again and again, 179
  Golden Prediction, 153
  Good luck or bad?, 81
  Is that your final answer?, 23
  MacArthur's magic number 115, 179
  Mind Reading, 115
  Mystery of 24 and Prime Numbers, 153
  Not quite 1 million, 46
  Phone Number Trick, 81
  Secret of 73, 81
  Superfast Additon, 179
  The Birthday display, 23
  The Count is IN!, 46
  Words, Words, & Breakfast, 46
Cardinal number, 6
Casting Out Nines, 12
Celsius to Fahrenheit
  Conversion, Fastest, 182
  Conversion, Traditional, 181
Check Digit Math, 11
  Casting Out Nines, 12
Check digits
  Review - Addition, 35
Check Digits
  Addition, 138
  Addition, 12

Creating, 11, 138
Fix Error in Multiplication, 140
Multiplication, 139
 Fix Error, 140
 Mod 7 Residue, 142
Non-Positional Number, 10
Positional Number, 10
Subtraction, 35, 138
Subtraction, 36
Test comparison check digit, 13
Understanding, 11
Commutative Property, 59
compatible numbers, work with, 31
Compensation, Using, 33
Computers
 ASCII, 151
 Base 2, 147
 Binary, Base 2, 148
 Bit & Byte, 148
 Convert Base 2 to Base 10, 149
 Math Behind Them, 145
Connect the dots, 89
Contour Figures Illusions, 163
Cost of a Baseball and Glove Problem, 197
Counting Numbers, 6
Counting PARTS up and down, 30
Creating a money shirt, 52
Cupids Arrow, 119
Curious methods for Multiplication, 107
Cyclic Numbers, 44

D

da Pisa, Leonardo. *See* Fibonacci
da Vinci, Leonardo, 125, 135
de Fontenelle, Bernard, 92
Descartes, Rene, 96
Distributive Property, 61
Divide
 by 2 (in half), 16
 Evenly by 11, 159
 Evenly by 2 - 10, 116
Divine Proportion. *See* Golden Section, *See* Fibonacci Golden Ratio
Divisibility Rules
 Even divisibility by 2 through 10, 116
Douchette, Jason, 39

E

EDUCATION = Problems!, 124
Einstein, Albert, 28
Endgame for the Day, 25, 52, 89, 119, 160, 188
Enormous numbers, 94

time of Archimedes, 94
Eratosthenes, 66
Euler's constant, 9
Exponentiation
 Negative Powers of a number, 177
 Positive Powers of a number, 177
 Rules of and Algebra, 178
 What is it?, 177

F

Factoring
 Prime Birthday Cake method, 68
Fahrenheit to Celsius
 Conversion, Fastest, 182
 Conversion, Traditional, 180
Fibonacci
 Architecture, 137
 Art, 135
 Golden Raio, 132
 Golden Section, 132
 His Books, 126
 His rabbits, 126
 History of, 125
 Human body
  Fingers, 132
 Human Design, 134
 In Nature, 128
 Numbers, Flowers & Plants, 129
 Phi/phi, 132
 Rectangles/Squares, 128
 Series - numbers - sequence, 127
 Spirals, 130
 Spirals & Sea Shells, 129
Fibonacci Series, 125
Finding the 6 in a bed of 9s, 196
Five Easy Ideas in Math, 30
Fractions, 67
 Add different denominators, 78
 Adding, 77
 Dividing, 80
 Least Common Denominator, 67
 Multiplying, 79
 Subtract different denominators, 79
 Subtracting, 78
 What are they, 77
Fraser's Spiral, 163
Fun with Numbers, 186

G

Gelosia Method Multiplication, 107, *See* Lattice Method Multiplication
Get off the Earth puzzle!, 25
gnuplot, 175
Golden Mean. *See* Fibonacci Golden Ratio
Golden Number. *See* Fibonacci Golden Ratio
Golden Pentagrams, 136
Golden Ratio
   Fibonacci, 132
Golden section
   Fibonacci, 132
Golden Section
   Art, 135
   Human Body, 134
Golden Triangles, 136
Gomutrika Method Multiplication, 108, 111
Googol, 97
Googolplex, 98
Graham, Ronald, 100
Graham's number, 99
Greatest Common Factor. *See* Least Common Denominator
   to find LCD/LCM, 70
   Using Upside-down Birthday Cake, 71
   Using Venn diagrams to solve, 73
   What is, 68
Gudder, Stan, 2

H

Hardy, Godfrey H., 122
Henderson, Chris
   Möbius card (2002 Games Magazine), 176
Hieron, King, 94
Hilbert, David, 92
Hindu-Arabic System, 147
Holy Family
   Golden Pentagram, 136
How much is 2 times 2?, 123

I

Impossible Shapes, 164
Infinitesimals, 95
Integers, 8
Irrational Numbers, 9
Irvin, Tim, 39

J

Jacobi, Carl, 56

K

Kasner and Newman, 97
Kasner, Edward, 97
Kelley, Colin
   gnuplot, 175
Kettering, Charles, 122
Klein, Felix
   Klein Bottle, 173
Kline, Morris, 168
Knuth's Up-Arrow Notation, 100

L

Large Numbers. *See* Enormous numbers
Lattice Method Multiplication, 107, 110
Laws of Math, 58
   Associative Property, 58
   Commutative Property, 59
   Distributive Property, 61
   Operations, 58
   Order of Precedence, 61
Least Common Denominator
   Finding, 67
   Upside-down Birthday Cake method, 70
   Using GCF to find, 70
   Using Upside-down Birthday Cake, 72
   What is GCF, 68
Least Common Denominator (LCD), 67
Leibniz, Gottfried, 148
Liber abaci. *See* Fibonacci
Line-crossing Method Multiplication, 111
Listing, Johann Benedict, 171
Locke, John, 122
Logic
   and Common Sense, 48
   Average Problems, 50, 82
   Definition, 48
   Difficult Problems, 51, 83
   Easy Problems, 49
   Reasoning out words, 47
   Strategies to solve, 49
Logical Methods of Math, 47, 82, 102

M

M, Heart, 8 Problem, 198
Math
    Laws of, 58
Math is Logic, 47, 82, 102
Math over the Years, 93
Mathematical Curiosities, 38, 65, 94, 125, **171**
Mathematical Properties, 58
Maxwell, James C., 102
Mesoamerican Long Count calendar, 146
method of exhaustion, 95
Michelangelo, 136
Möbius
    Card, 176
    Create a strip, 171
    Create Klein Bottle, 174
    Create Möbius Card, 176
    Four-dimensional Object, 173
    Impossible Shapes, 171
    Klein Bottle, 173
    Universal Recycling Symbol, 173
    Work with strip, 172
Möbius, August F, 171
Mod 7 Residue Method, 142
    Working with, 143
Mona Lisa
    Fibonacci Rectangles, 135
Mond Crucifixion
    Golden Triangle, 136
Money Shirt, 52
Morse, Harold M, 92
Moscow papyrus, 108
Moser, Leo, 100
Moser's number, 100
Müller-Lyer Illusion, 161
Multiplication
    Brother Juan Diez, 108
    Brother Juan Diez Method, 113
    Gomutrika Method, 108, 111
    History of, 107
    Lattice Method, 107
    Line-crossing Method, 111
Multiplication Methods
    Historical methods, 107
Multiplication Problems
    Check Digits, 139
    Fix Errors with Check Digits, 140
    Mod 7 Residue, 142
Multiply
        number by series of 9s, 41
    12's tables using 1/2 Fractions ratio, 183
    99 by any number less 100, 87
    any number by 11, 19
    any number by 12, 84
    any number by 9, 40
    by 10, 15, 156
    by 2, 15, 156
    by 2 or power of 2, 42
    by 5, 17, 42, 156
    by powers of 2 – 2, 4, 8, 22
    Mixing it Up, 157
    Number of digits by same number 9s, 40, 117
    Squaring
        Any number 1 or 2 above/below a previous
            square, 183
        Any number end in 0, 86
        Any number end in 5, 86
        Numbers near units of 10 or 100, 185
    Three digits by 1001, 88
    Three digits by 101, 185
    two digit number by two digit number, 21
    two digit numbers different by 2, 184
    Two digits by 101, 87, 184
    two digits by 11, 18
    two number less than near 100, 85
    Two numbers with special relationsip, 184
    Two teen number differ by 1, 88
Murphy's Law, 57
myriad, 95

N

Neurological Tests, 196
Newton, Isaac, 148
Nominal Numbers, 6
Non-orientable
    Topology, 171
Non-Positional Number, 10
Notation
    Knuth's Up-Arrow, 100
Number Oddities
    13 and 15 & Raising by summation, 186
    5 twos and math operations, 187
    81 and 512 & powers, 186
    Apocylyptic 666, 43
    Billion a billion, 43
    Cyclic Numbers, 44
    Digital delight from 1 to 9, 186
    Narcissistic Numbers, 186
    Reversal of multiplication, 186
    Symmetry of number calculations, 187
    XI+I=X, 45
Numbers
    aleph nought, 6
    Place Value Notation, 7

Review of Types of Numbers, 5
What are numbers, 5
Numbers, Type of, 6
  Algebraic Numbers, 9
  Cardinal number, 6
  Counting Numbers, 6
  Integers, 8
  Irrational Numbers, 9
  Nominal Numbers, 6
  Ordinal Numbers, 6
  Rational Numbers, 8
  Transcendental Numbers, 9
  Whole Numbers, 6
Numeral Systems
  System of Numeration, 145

O

Operations
  Definition, 58
Optical Illusion
  A few examples, 161
  Natural Examples, 160
  Science Behind them, 160
Optical Illusions, 160
  Black Dots, 165
  Contour Figures Illusions, 163
  Fraser's Spiral, 163
  Gray Center bar, 164
  Impossible Shapes, 164
  Mathematics, 160
  Müller-Lyer Illusion, 161
  Orbison's Illusion, 162
  Poggendorff Illusion, 162
  Three dimensional World, 165
  Vertical-Horizontal Illusion, 161
  Zöllner Illusion, 162
Orbison's Illusion, 162
Order of Operations, 61
Order of Precedence, 61
  Make a difference, 64
  PEMDAS, 62
  Why we need, 62
Ordinal Numbers, 6

P

P(E) Probability. *See* Probability
Palindrome Numbers
  Create from 3 digit number, 38
  World of, 38
Parthenon

Fibonacci, 137
PEMDAS, 62
  Acronym for, 63
Phi/phi
  Fibonacci, 132
  What are they, 133
Place Value Notation, 7
Plato, 2, 122
Plutarch
  Mestrius Plutarchus, 56
Plutarchus, Mestrius. *See* Plutarch
Poggendorff Illusion, 162
Positional Number, 10
Prime Birthday Cake factoring, 68
Prime Numbers, 65
Primes
  List of 1 through 2000, 155
Principle of buoyancy, 95
Probability, 102
  Coin Toss, 102
  Coin Toss, 3 coins, 105
  Formula, 104
  History of, 103
  Marble from a bag, 104
Probility
  two child gender, 106
Problems
  Easy Logic, 49
  Logic - Difficult, 51
  Logic average, 50
Process of 11
  Addition, 159
Proclus, Lycaeus, 28
Properties
  Mathematics, 58

Q

Quotations of the Day, 2, 28, 56, 92, 122, 168

R

Ramsey theory, 100
Raphael, 136
Rational Numbers, 8
Recycling
  Symbol & Möbius, 173
Rhind, A. Henry, 108
Rhine Papyrus, 108
Riemann hypothesis, 99
Riemann, Bernhard, 99
Rules of Exponentiation
  Negative Rule, 178
  PowerRule, 178

Product Rule, 178
Rules of 1, 178
Zero Rule, 178

S

Sagan, Carl, 98
Sam Loyd, 25
Sand Reckoner, 95
Scientific notation, 95
    Naming of, 96
    Powers of 10, 97
Sequential Series of Numbers
    Summing, 118
Seurat, Georges, 137
Sieve of Eratosthenes, 66
Simple Math is All You Need, 14, 40, 84, 116, 156, 180
Sirotta, Milton, 97
Skewes' number, 98
Squaring
    Any number between 10 and 120, 158
    Any number end in 0, 86
    Any number end in 5, 86
    Any number one above/below previous square, 158
    Any number two above/below previous square, 158
Stienhaus's polygon notation, 100
Subtraction through Addition, 192
Sumario Compendioso. See Brother Juan Diez
    Method Multiplication, See
    Multiplication:Brother Juan Diez
Summing
    Sequential Series of Numbers, 118
System of Numeration, 145

T

Temperature
    Algebraic formula for F to C & C to F, 182
    Converting Fahrenheit to Celsius and back, 180
Thinking Algebraically, 197
Topology
    Listing, Johann B, 171
    Non-orientable, 171
Trachtenberg Method, 107, 109
    any number by 12, 84
    Multiply by 11, 19
Trachtenberg, Jakow, 109

Transcendental Numbers, 9
Numbers
    Natural Numbers, 6

U

Unary System
    Base-1, 145
Upside-down Birthday Cake
    finding GCF, 71
    Finding GCF and LCD, 70
    Finding LCD, 72

V

van Landingham, Wade, 39
Vedic Math
    Multiply any by 11 through 19, 85
    Multiply digits by 9s, 40
Venn diagram
    Solving for LCD/LCM, 75
Venn Diagram
    finding GCF and LCD, 73
Venn diagrams
    Solving GCF, 73
Vertical – Horizontal Illusion, 161
Vitruvius, Marcus, 94
von Neumann, John L, 168

W

Walker, John, 39
Whole Numbers, 6
Williams, Thomas
    gnuplot, 175

Z

zephirum, 125
Zero, 7, 10
    Babylonians, 146
    Brahmagpta, 146
    Importance, 146
    Mayan concept, 146
    Special Role, 7
Zöllner Illusion, 162

www.ingramcontent.com/pod-product-compliance
Lightning Source LLC
Chambersburg PA
CBHW081113170526
45165CB00008B/2442